MW01482787

How to Get Hired in Law Enforcement

Police and Fire Publishing
1800 N. Bristol, Suite C408
Santa Ana, CA 92706

Email: Bolly7@aol.com
Website: policeandfirepublishing.com
ISBN: 978-0-9821157-3-2

Note to the Reader: Concepts, principles, techniques, and opinions presented in this manual are provided as possible considerations. The opinions, application, use, or adoption of any concepts, principles or techniques contained in this manual is to be used at the discretion of the reader and/or organization.

How to Get Hired in Law Enforcement
3rd Edition

Authors Pete Bollinger and Steve Winston
DVD available at hiredbypolice.com

HOW TO GET HIRED IN LAW ENFORCEMENT AND FIRE

HOW TO GET HIRED IN LAW ENFORCEMENT AND FIRE
(DVD)

The Authors

Pete Bollinger first became interested in law enforcement while serving in the United States Air Force where he attained the rank of Sergeant. While in the military, he obtained AA Degrees in Industrial Security from the College of the Air Force and Criminal Justice from Devils Lake College in North Dakota. After an honorable discharge, Mr. Bollinger continued his education, obtaining a Bachelor of Science Degree in Public Administration from the University of La Verne and a Master's Degree in Management from the University of Phoenix.

Pete Bollinger has been a police officer with the Santa Ana Police Department since 1988 and has worked a variety of assignments throughout his career. First and foremost, he is proud that the majority of his work history was on patrol. He also served in an undercover capacity in narcotics and vice programs. Mr. Bollinger was fortunate enough to earn an assignment in Personnel, specifically in the Background Investigation Unit. He promoted to the rank of Sergeant, supervised the Felony Case Filing Unit and currently supervises patrol officers.

Steve Winston, Bollinger's co-author, mentor and good friend, has been a Santa Ana Police Officer for over 28 years, working patrol, foot patrol, task force, and 11 years as a Motor Officer where he designed and presented a highly recognized bike safety program to over 230,000 students. Steve has been a Background Investigator for 12 years and has completed over 1,000 background investigations for both law enforcement and firefighter positions. He is highly regarded in his field and sought after to give presentations at academies and colleges throughout the United States. Steve is currently Vice-President of the California Background Investigator's Association.

The authors have nearly five decades of collective experience in law enforcement and have created the most comprehensive book on the realities of getting hired in public safety that exists. They have conducted thousands of background interviews and you will be the beneficiary of their experience and candor as they take you through the hiring process in this easy to understand and extremely informative book. If public safety is your chosen career path, look no further - - this book is your ticket to success.

IN HONOR OF...

Yes, do mourn for the officers and firefighters who have given their life in the line of duty, but know they did this with a willing heart. In reverence to their memory and for the families of these brave men and women, a percentage of every book sold will be donated to a police and firefighter widow's and orphan's fund.

Teddy Roosevelt is quoted in a speech from the early 1900's that is truly understood by police officers and firefighters.

"It is not the critic who counts; it's not the man who points out how the strong man stumbled... Credit belongs to the man who is actually in the arena, whose face is marred by dust, sweat and blood; who strives valiantly; who errs, to come short and short again, because there is no effort without error and shortcoming; it is the man who actually strives to do the deeds; who knows great enthusiasms and knows the great devotion; who spends himself in a worthy cause; who at best knows in the end the triumph of great achievement. And who at worst, if he fails, at least fails while daring greatly, so that his place shall never be with those cold and timid souls who neither know victory nor defeat."

If the above passage "speaks" to you then you probably have the heart to become one of us.

DEDICATION

This book is dedicated to all Police Officers and Firefighters who follow their calling despite the obvious dangers; particularly to those who have lost their lives protecting us. They live on in our hearts.

Pete Bollinger
A special thanks to my wife Carmen, who encourages me to follow my dreams, and to my son Matthew and daughter Chelsea; I couldn't have chosen two better kids or anyone I love more. To all of my friends who I never quite let get too close to me: Jane, Skip, Debbie, Brenda, Natalie, Elva, Israel, Jim, Bo, Brian, Ernie, Ron, Trisha, Sergio, Silvio, Dave, Arlene, Tina, Tara and all of my brothers and sisters in public safety.

Steve Winston
The support of my wife Jane, sons Ben (a police officer) and Steve, and daughters Shauna and Taylor have continued through the years to be the inspiration to answer my calling - to help others enter the field of law enforcement that has made my life so complete.

Table of Contents

Foreword

This is the most comprehensive book you will find regarding public safety hiring procedures. We have compiled anecdotal stories and advice from leading experts in the field, in an easy to follow format that will be your ticket to a career in law enforcement or firefighting.

This book is intended for young men and women who aspire to be Police Officers, Deputies, Firefighters, Crime Scene Investigators, Dispatchers, Arson Investigators, Security Officers, government officials, civilian law enforcement positions, and any high profile career. Family members and friends of those aspiring to be in public safety can also benefit from this book for a better understanding of what is required by their loved ones; they will be amazed at what these professionals go through for the honor of serving the public.

Police and fire administrators are encouraged to read this book as an overview of the hiring process from an insider's point of view. It will prove to be beneficial in providing a better understanding of the investigation process as relates to hiring quality candidates, and it may also even help streamline this process throughout the country. Agencies need to have a long-term outlook rather than meet short-sighted objectives. Departments who have hired substandard employees wish they had spent what was necessary up front to discover problems with prospective employees. Every organization should consider allocating a higher percentage of their budget on more background investigators and/or overtime as a form of risk management; it would more than pay for itself over the course of time. Support your background investigators and allow them time to follow leads with minimal time constraints. Your only edict and direction should be to "hire only the best" ethical applicants with high moral character.

Case in Point: The hiring of an officer by one of the largest police departments in the country. The officer was the leading man in a public relations nightmare the media labeled a scandal. He went to prison for planting evidence, lying under oath and maliciously shooting a man he knew was unarmed. The department paid out millions of dollars in settlements and released dozens of other arrestees due to that officer's unethical and illegal activity.

A little known fact is that the officer was actually disqualified by another police department prior to his employment. At the time, this department was hiring numerous officers within a very short time period, which, no doubt, placed a huge amount of pressure on the background investigators to produce higher numbers. It may have led to honest mistakes by overwhelmed, well-intentioned investigators.

How to Get Hired in Law Enforcement and Firefighting

Introduction

If the title of this book caught your eye you are probably interested in becoming a police officer, firefighter or any position in public safety. You are concerned that you live in a society where rules are not followed and laws are not obeyed. You are different; you enjoy structure and feel that laws should be enforced. You are uncomfortable around people who participate in criminal activity and believe they should be held accountable for their actions. You have the desire to serve and protect the public and have a positive impact on society! This book will help you conquer your personal roadblocks to entering public safety. To put it confidently and truthfully, this book will change your life!

Both of the authors have extensive experience in hiring police officers, firefighters and civilian law enforcement positions; they will share valuable information to make you a better applicant and give you an edge over other candidates in that competitive process.

There are literally thousands of sworn and non sworn positions in public safety and preparing to pass the hiring process can only be beneficial to you and your career aspirations. This book will assist you in developing a public safety mindset which is vital to your success.

Below are some of the positions that have similar requirements in the hiring process as Police Officers and Firefighters.

Police and Fire Dispatcher

Harbor Patrol

Canine Security Officer

Parking Control Officer

Security Officer

Animal Control

Police and fire departments are bonded by a sense of brotherhood in many ways, mainly because they are on the front lines together and have seen the same horrors and lived through similar experiences. Police and fire departments are becoming even closer as societal demands and expectations are continuously increasing for both professions.

Background investigations were virtually non-existent for fire personnel and civilian law enforcement positions ten years ago, but are now commonplace. There is also a push to establish polygraph and psychological examination requirements for these applicants. Many police and fire departments utilize the same background investigators for all of these professions.

This book contains advice from leading experts in the field and tidbits from actual police and fire supervisors who regularly sit on interview panels. These experts will explain what they are looking for and let us into their world regarding some of the best and worst answers they have heard from prospective applicants. The first important tip to remember is **"the prepared applicant is the one who gets hired"**.

Choosing Public Safety

Is public safety right for you? Before entering this career field, you must consider if the danger involved fits your personality. Please do not be offended if you are not made for this type of work, very few are! Most in public service feel they have a "calling," not necessarily in the religious sense, but an intuitive unexplainable desire to serve in public safety. If you do not feel drawn to serve the public and are only interested in fringe benefits (status, power and salary), you will find yourself in trouble in a variety of ways we will cover in detail later in this book. Only serious applicants need apply; the opposite could result in death or serious bodily injury to you, your partner, or the public.

Passion for the Job

For those of you who have a desire to enter public safety you should first understand that you are unique individuals. The intrinsic need to be in this profession is so strong it may be difficult for you to articulate why. You will need to put into words why you should join those entrusted to protect and serve the public, and we will assist you in doing so.

What Are the Dangers of Public Safety?

Law enforcement is a dangerous profession. Most arrestees will be compliant, but a few will fight even after being placed in handcuffs. If you are afraid of physical contact, easily intimidated or weak in resolve, consider an alternative career.

13

It is exciting and "cool" to be a cop or firefighter and to also carry around the status of these professions off duty; however, you must ask yourself, "am I willing to pay the ultimate price for this privilege?" Will you rush into a burning building to save people even if there is a possibility of being consumed by the fire? Will you confront a violent parolee who says he won't go back to prison? These situations are very real and not glamorous like the movies make them out to be. Are you willing to die for a misdemeanor crime and then be second guessed by the critics after your death? Are you willing to expose yourself to toxins from a chemical fire that could result in an early death?

Decontamination Unit

You will constantly be in the presence of narcotics and weapons, even when you are not aware of them.

Have you considered other dangers involved in the area of public safety? Getting shot as a police officer or being burned as a firefighter are possibilities, but a more common threat is from blood borne pathogens, exposure to viruses such as meningitis, hepatitis, HIV/AIDS, and many other illnesses, some of which in the prisons and jails have yet to be identified. You may get sick more often than the average person due to these types of exposures.

Searching vehicles is a common reason officers get stuck with needles. Many IV drug users are infected with some type of disease. Due to privacy laws and other variables, testing the owner of the hypodermic needle/syringe is impractical; officers must simply "hope" they were not infected.

As a firefighter, there will be substantially more risks from hazardous materials and toxic waste than from any other job. Yes, you will wear self-contained breathing apparatus tanks; however, this does not stop particulate matter involving carcinogens no matter what the so-called experts tell you. There are numerous documented and undocumented stories of firefighters whose lives have been cut short by exposure to hazardous chemicals. Firefighters also have a higher percentage of cancer than the average person for obvious reasons.

As a police officer, the odds of you getting shot are rare; however, you will be in fights, get soaked with the body fluids of unclean people and occasionally be injured. You will most likely die years earlier than the average person, due to the continuous acceleration and deceleration of your heart rate, blood-pressure, and stress the job takes on your body. If you truly have a calling to be in public safety, these obstacles will not deter you.

Threat of Jail

Public Safety is one of the only careers in which you can go to prison for trying to do the right thing. The current trend is to prosecute officers for their actions when attempting to take a violent suspect into custody. This usually occurs after a high speed pursuit. The public, including the district attorney's office, believe they can duplicate the speed and stress of an actual situation using a videotape of an incident. Every split second is scrutinized and evaluated as if the officer had the time to consider all of the options the Monday morning quarterbacks deem appropriate. It can be explained like this: Go to a batting cage and as the machine is about to throw a 90-mile-per-hour pitch, look up; the speed will seem unbelievable, but watch it for a few minutes and realize how much slower it looks. That is what is happening with videotape evidence and some of the critical public who actually believe they could do our job.

Everyone has a video/cell phone

Unfortunately, they will not understand the hundreds of thoughts going through an officer's mind and the safety issues he is concerned with. You must accept that you are held to a higher standard; standards which could not be met by the attorneys prosecuting you or the judge sentencing you, and standards raised to an unattainable level by the public that vilifies you. If you accept this fact, you will enter into a world that is only understood by police officers and firefighters.

The Media

The media and court system love to create controversy and put police officers in harm's way. Why, you may ask. The answer is simple; anything that will sell papers and/or make the evening news a little bit more interesting is fair game. It is ignorant for people to believe that information from the newspaper or evening news is all true. We have been on too many scenes where we knew the truth firsthand, but hearing the reporter later, made us wonder if they were talking about the same incident! We simply shake our heads at the inaccuracies and outright lies the public is often told. It is not uncommon to watch a criminal who you would not want living in your neighborhood evolve in the media from a dangerous parolee who fled in a vehicle at over 100 miles an hour to an "innocent motorist" in a matter of weeks. Officers and firefighters just grimace, shake their collective heads and move on.

Do not believe everything you read

If you think the media and the public will appreciate you putting your life on the line every day, think again. Throughout history, officers have been second guessed and persecuted, and today is no different. Over a hundred years ago, Wyatt Earp and other deputies tried to disarm outlaws carrying weapons within city limits. It led to the famous shooting known as the "gunfight at the OK Corral." What is not well known is that Wyatt Earp and his deputies were put on trial and Earp had to mortgage his house to pay for his legal defense. Not much has changed.

America has a tendency to throw common sense out the window and root for the underdogs. As a police officer you will never be the underdog. It never ceases to amaze how otherwise intelligent people can be duped by defense attorneys and the media into believing a defendant is innocent. Innocent people do not go to prison. Our system is in such a mess that it takes several arrests and a negative probation report to send someone to the big house.

Threat of Lawsuits

Firefighters are not immune from being second guessed. There is a new industry of mercenary type retired firefighters and paramedics (usually a medically suspicious retirement) who will testify for any plaintiff that meets their fee. Most lawsuits will derive from the response, deployment and decision making at structure fires. The trend over the last few years has been the public making derogatory comments about firefighters. View the local editorial section on a regular basis and you will read

18

negative comments from all segments of society regarding firefighters. This was once considered unfathomable, but in today's society is not surprising. Police officers have unfortunately grown accustomed to this treatment, but it is new to fire agencies. New breeds of attorneys are attacking first responder firefighters due to the deep pockets of the agency. These parasites will question you on your response to the scene, Did you make a wrong turn? How fast were you going? What were you doing prior to the call, and did that delay your response? We drove that same route and got there three minutes faster without the use of lights and siren. How do you explain that, sir? The court system that in the past has given an unwritten respectful immunity regarding litigation to firefighters is changing and you need to be prepared for the possibility of lawsuits coming your way.

Are you prepared to defend every action or inaction on video as you arrived on scene in court? Everything is on video these days; cameras, cell phones and the old fashioned video camera peering out of a window. Did you administer too much narcan (medication) or not enough? Did you start cardio-pulmonary resuscitation (CPR) soon enough and did you do it right? Plaintiff's attorney to the firefighter, Tell me the procedures of CPR in detail. Could you have entered that engulfed structure fire and saved those people? Family members who have not seen the homeless heroin user for twenty years will show up in black and sue you for their "grief." They will allege improper medical treatment and had you used proper procedures he wouldn't have died. It is your fault! If you are thinking it won't happen, you're wrong; it definitely could.

If reading this chapter creates doubt about your "calling," there is no disgrace in acknowledging that public safety may not be the career for you. However, as you will learn later in this book, public safety is comprised of many different jobs, including civilian positions and one of these may be your "calling".

Salary, Benefits and Retirement

There are many benefits to a career in Public Safety; primarily job security and a nice income. Flex scheduling is very common and you will most likely be working a 3-12 schedule (three 12-hour shifts per week) or a 4-10 schedule (four 10-hour shifts per week). You will be well taken care of in retirement as most departments offer what is known as the 3@50 plan; accumulating three percent of their annual salary per year of service and can collect at 50 years of age. There are many ranges in pay. The lowest paid department now may be the highest paid in a few years and vice versa. You need to be prepared to work weekends, holidays and odd hours. Your family needs to buy into this because it is part of the job. After a few years you may have enough seniority to get a Saturday or Sunday off and/or work dayshift.

Family Preparation for Entering Public Safety

Wives, husbands, mothers and fathers were interviewed and asked how they felt when they first heard about their loved ones going into the field of law enforcement and firefighting. Many responded that they are proud but nervous. Most were not surprised their loved one was entering a career in public safety stating they saw something in the applicant's youth that headed them in that direction.

There is a difference between those who met their spouse when they were already officers and those who met their spouse before they decided to become an officer. Couples who came together after the officer was already working in law enforcement were more accepting of the dangers, the hours and the police culture than their counterparts. Spouses or fiancées of those just becoming officers described feelings of fear and concern for their well-being. Most describe their efforts to dissuade the officers from entering public service as futile.

While recruiting at fairs and expos, we constantly hear people reminiscing with comments such as, "I was going to be a cop, but my parents didn't want me to;" "I was going to be a firefighter, but my wife said it was too dangerous;" "My husband didn't want me working shift work;" "My girlfriend said she wouldn't stay with me if I did this." If comments like these kept these people from pursuing their dream, they were either very weak or perhaps it was not a strong enough dream! If someone can dissuade you from applying, very likely you are not passionate enough for a career in public safety. It is a calling that cannot be pushed aside if it is **meant to be**.

Non sworn and/or civilian law enforcement positions

There are thousands of law enforcement positions other than a police officer. We will discuss many of these non sworn and civilian positions and give real world definitions of each of them. Police officers are sworn to uphold the United States Constitution 24 hours a day, 365 days a year. Any position other than sworn at a police department is called "non sworn" and are considered civilian law enforcement employees. These are noble and worthy endeavors also assisting in public safety.

The most common method of becoming a sworn officer is to obtain a position as a security officer and/or loss prevention officer. These gateway jobs can be beneficial in determining whether traditional law enforcement is right for you or if you should pursue the civilian side of the profession.

Security Officer
And/or
Loss Prevention Officer

One of the most common positions an individual applies for on their way to a law enforcement position; in fact one could argue it is a law enforcement position. Many security officers choose to stay at their company for an entire career and enjoy promotions, status, supervisory experience and job security. A store will either contract with a security company or hire their own loss prevention officers. Being an employee of a chain of stores can offer unlimited opportunity. Most monetary loss is attributed to employee theft and will give you an opportunity to be a detective and put together cases for prosecution.

Requirements for this position:

Good physical condition – You will be taking people into custody; most of them will be cooperative, but occasionally you will encounter a fighter. When this happens you may not have back up and do not have the civil or criminal protection afforded police officers. However, you may retreat and call police if the situation is spiraling out of control.

Intelligent –Quickly discern situations and understand when and how to go to the next level.

Self-control – There will be times when you are angry, frustrated and disrespected. Your discipline will be tested.

High ethics and character – Your honor and integrity should be above reproach. Testimony in court is part of the position and the majority of the time it will be a "he said, she said" situation. The judge and jury must trust you in order for the prosecution to obtain a conviction.

Keep in mind this position incurs a high amount of liability, both civil and criminal.

Private Investigator

Duties include:

Surveillance

- Utilize state-of-the-art technology in order to provide quality video documentation.
- Spot check(s) of claimant's residence to determine daily activities.
- Neighborhood or area investigation/interviews.

Workers compensation

Fraudulent claims are rampant throughout the system. You may be assigned to conduct surveillance and capture movement on video. Stories of employees claiming a back problem and later videotaped dancing or surfing are not uncommon.

- Provide photographs, diagrams and/or observations documenting the accident/injury site.
- Interview employers, co-workers, or other pertinent witnesses regarding the accident/injury in question.
- Obtain official documentation as appropriate that relates to the accident/injury investigation.

Subpoena investigation

- Locate claimants, witnesses, or other necessary persons to facilitate interview, and document signing or subpoena service.

Health provider investigation

- Determine if billing reflects actual care given.
- Video surveillance on all clinic patients & employees documenting duration of visit.

Mortgage fraud

- Locate and interview borrower(s) and employer(s).
- Re-verify employment and/or assets with written re-verifications
- Have Borrower sign IRS Form 4506 or get a VOE from employer
- Verification of Deposit (VOD)
- Investigate employment, occupancy, front buyer, straw buyer, down payment/source of funds, undisclosed properties/mortgages, etc.

Other types of investigations

- Face-to-face contact with claimants for signing of documents such as "Agreed Statement of Facts" or answering specific questions/interrogatories.
- Taking recorded statements in those instances in which the claimant is unresponsive to specialist's efforts to obtain information by telephone.
- Testifying in court.

There are numerous positions in law enforcement that do not require an individual to face danger; these positions can be quite satisfying and contribute to both law enforcement and society. Below is a list of some of these positions:

Affirmative-Action Specialist

Background Investigator

Cold Case Investigator

Correctional Records Specialist

Court Liaison Officer

Computer Specialist

Coplink Specialist

Crime Lab Technician

Crime Prevention Officer

Crime Research Analyst

DARE Officer

Dispatch 800 Megahertz Technician

Document Examiner

Electronic Security Technician

Event Security

Executive Police Secretary

Fingerprint Examiner

Forensic Ballistics Examiner

Firearms Examiner

Human Resources Specialist

Intelligence Analyst

Judicial Assistant

Law Enforcement Mechanic

Law Enforcement Motorcycle Mechanic

Law Enforcement Products Developer

Legal Secretary

Management Aide

Management Analyst

Manpower Allocation Specialist

Personnel Services Specialist

Police Cadet

Police Records

Process Server

Programmer Analyst

Systems Technician

Microsystems Specialist

Motorola Radio Technician

Narcotics Treatment Program Coordinator

Office Specialist

Police Investigative Specialist

Property and Evidence Technician

Range Master

Restitution Specialist

Supply Clerk

Uniform Store

Victim/Witness Advocates

Vocational and Rehabilitation Specialist

All of these positions contribute to law enforcement and benefit the public. There are many other ways to assist law enforcement and also make a very good living. Follow your passion and match it with research and you will ultimately land your dream job!

Additional non-traditional law enforcement related positions

Bail Bonds

Most people arrested are offered bail. It is important that the bail bond representative understand both sides of the law.

Civilian Jailer

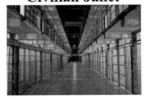

Civil Rights Analyst Commissioner

Computer Services

Court reporter

Courtroom testimony or any other type of courtroom activity cannot happen without a court reporter. Going through a technical school to learn how to utilize shorthand on s specific machine is mandatory. It is not uncommon for a court reporter to make $100,000 a year.

Civilian Bailiff

Depending on the requirements of the state, county or even the judge the bailiff may be a civilian. Limited police powers while on duty only, saving the organization a tremendous amount of money over the more expensive sworn personnel.

31

Crime Artists

Crossing Guard

Data Processing

Generally an entry level position, but can lead to other opportunities within the criminal justice system. Necessary skills are typing and attention to detail. The fantastic fact about data processing positions is that they are abundant and allow you a foot in the door at the organization of your dreams.

Discovery Clerk; prepares evidence for court and acts as the liaison between police departments and the district attorney. Will develop close relationships with detectives and prosecution sections from municipal and county police.

Evidence Technician; almost everything in court is challenged, especially what is referred to as "chain of custody". Most departments have professionals that specifically process, preserve and prepare evidence for court.

Filing Clerk

Fiscal Officer; law enforcement has huge budgets; millions of dollars that need to be accounted for. Computing different schedules, overtime and issuing payments to vendors. This is a challenging, but rewarding position.

Funeral Processions

Halfway House Worker

Harbor Patrol

Hazardous Materials

Helicopter Pilot

Highway Safety Specialist; you have all driven past the weigh scales on the freeway. The Department of Transportation and Safety has numerous positions available and the work you do will be saving lives.

Home Test Kits

Internal Revenue Service Employee

Law Enforcement Reporter

Law Firm Employee; there are numerous positions available in all types of law firms. Litigation is alive and well in America and the opportunities to prosper in this area are as well. Areas of law are: Lawyer, law clerk, paralegal, legal researcher,

Law Library

Mentally Retarded Advocates

Office Manager

Court positions are plentiful and performing at a high level will lead to numerous opportunities.

On School Campus

On-Campus Security which is comprised of numerous positions within school district

Polygraph Examiner

Extensive school and training program developing a thorough understanding of behavior analysis.

Police Photographer/Videographer; if you enjoy law enforcement and photography/videography equally there are many opportunities in this career field. Duties may range from a police yearbook to videotaping a confession of a murder suspect. You will have a creative opportunity when producing training tapes and sound-bites for the media.

Phlebotomist

Process Server

Social Worker

This photograph needs no explanation.

Subpoena clerk; works with court personnel and coordinates court appearances by victims, witnesses and police officers. This position requires flexibility and the ability to plan and anticipate.

Tow Truck Drivers

Traffic Revenue Generator

Videographer

Police Departments

Dispatcher, Records, Civilian Investigator, Police Service Officer, Secretary, Filing Clerk, Subpoena Clerk, Animal Control, CSI.

Animal Control

There are three types of animal control officers at a municipal or county department.

1. The first type is an individual who is dedicated to the preservation and humane treatment of any type of animal.
2. Individuals who wish to someday be a police officer. This position will have you working side by side with patrol officers, making both misdemeanor and felony arrests.
3. An individual who is simply looking for a job. This person will not do well in this career field. If it is not a passion look for another position.

Community Service Officer

Another way to assist the community in a civilian capacity.

Crime Scene Investigator

Crime scene clean-up

An unbelievable opportunity for growth and millions of dollars is in the disposing and cleaning of hazardous materials. It could be a chemical spill from a gasoline truck or a messy crime scene. Laws have been implemented over the past decade to properly dispose of contaminants. Just a few short years ago a murder scene, including blood and other body tissue would simply be washed away with a firefighter's hose. Many times it would only disperse the matter into particles invisible to the eye; leaving vehicles and pedestrians to take the components home on their shoes. With newly discovered knowledge of environmental damage caused by improper disposal this industry has exploded.

Dispatcher

What is the difference between a police and fire dispatcher? Can you do both? 911 calls go to police first and if necessary passed on to fire. Some agencies require dual dispatching duties.

Parking Control Officer

This requires discipline and a thick skin; citizens will display anger, hatred and ridicule on a regular basis. The benefits to the community are significant; without these professionals every city would look like a war zone and/or third world country.

Park Ranger

Police and Fire Receptionist

Police and Fire Secretary

Range Master

Records and Computer File Technician

Serologist

Traffic analyst

Congratulations! If you are still reading this book, you just may have what it takes to succeed in public safety.

The Awakening...The "Aha" Moment

Most officers and firefighters nearing retirement age were indoctrinated into public safety by watching television. Back then, popular television shows portraying police officers and firefighters, such as Adam 12, Emergency, Hill Street Blues, Hawaii 5-0, and even Andy Griffith, piqued our interest and led us in the direction of having respect for authority and a desire to enter public safety.

Today's future public servant became interested for a variety of reasons such as television shows or movies like "Cops" or "Back Draft," and more practical reasons like salary and benefits, time off and vacation accrual. Considering quality of life issues when looking for employment is common from the current generation of young people; conversely, these questions were never asked nor even considered by past generations. It is not wrong to consider wages and benefits, but it should not be a primary reason to enter public safety.

For some of you, it is difficult to pinpoint why you are interested in this career field and articulate the reasons you would be good at it. Others can point to a specific incident in detail which led to their calling or "epiphany," defined as an intuitive grasp of reality through an event.

To give you an example of an epiphany, I will share my story. I was on the high school baseball team and we were playing an away game in a ghetto area. We finished our game early and were waiting on the bus for our tennis team to finish their match. One of the players on our team leaned out of the window and for no reason at all began yelling racial remarks at a Hispanic male who was in his vehicle picking someone up at the school. After a barrage of insults, the man was getting upset. He stepped out of his vehicle and walked onto the bus. He took a bat out of the equipment bag and challenged everyone on the bus to fight. The man was a parolee. He had a commanding presence and I was impressed with the way he manipulated the team with intimidation. He made the "racist idiot" look like a coward. I looked around and observed everyone polarized by fear. What surprised me was that I was not frightened; instead, I was planning what I would do if he started hitting my teammates with the bat. After the parolee left the bus, we droned a collective sigh of relief. I was curious as to why my mind continued to consider options during the incident; anyone else may have been petrified in this situation and I should have been. Then it hit me, I believed my calling was to be a police officer.

You may or may not have an incident such as this that lets you know what your calling is, but keep in mind if you are able to think and perform under stress in difficult situations and can logically and strategically process what your future actions will be if something goes wrong, then you are a strong candidate for public safety.

A colleague shared a different kind of awakening. His best friend growing up always wanted to be an officer and directed his life that way. He joined the Police Explorer Program, read everything he could about police officers and eventually did become an officer. My colleague had no intention of joining a police department and instead focused his attention toward obtaining a position in the medical field. After a year, for a reason he could not pinpoint, he changed his mind and decided to become a police officer. He thought he would be going to the academy a few months later. He was shocked when he realized it would not be that easy. It took him a few years of determined effort to finally get an academy spot and he is now a detective at a medium size agency. Though he did not have an "aha" moment, he was still somehow drawn to this career field. Had this book been available at the time, he would have understood the police culture and no doubt would have been hired much sooner.

Prior to my epiphany, I was considering a career in firefighting, but due to the incident in high school, I knew my calling was to become a police officer. My career choice was proven correct one evening on patrol when my partner and I came upon a residence in flames. We heard a baby crying in the house and entered through the garage. We were quickly overcome by smoke and were forced to retreat. It turns out the "crying baby" was actually a gas line making a high pitched noise from the flame. After this incident I knew I made the right decision to enter law enforcement. Firefighter friends have told me they think police officers are crazy confronting suspects with guns. I think they're crazy entering burning buildings! I think you understand what I am getting at, **"You don't choose a career…a career chooses you!"**

What is Required?

Job Dimensions Necessary for an Applicant to be Successful

Public safety officers must possess qualities that improve their probability of success. These traits are common and intrinsic with most police officers and firefighters. They are:

Communication Skills – The ability to express yourself clearly in writing and speech.

Problem-Solving Ability – The ability to size up a situation, identify a problem and make a logical decision.

Learning Ability – The ability to comprehend and retain factual information.

Judgment Under Pressure – The ability to make sound decisions under pressure.

Observation Skills – The awareness of surroundings and attention to detail.

Willingness to Confront Problems - The willingness to be assertive and confront a situation.

Interest in People – The willingness to understand and work with people.

Interpersonal Sensitivity - Resolving problems while showing empathy.

Desire for Self Improvement - Desire to seek and learn new knowledge.

Appearance – Personal and professional pride in one's looks.

Dependability - Accuracy and timeliness of work.

Physical Ability - Endurance and ability to meet the demands.

Integrity - Refusing to engage in or tacitly approve of unethical or illegal conduct.

Operation of Motor Vehicle - The ability to safely operate and control a vehicle.

Credibility in Court - The ability to testify without character impeachment.

Start at the top of the list and evaluate yourself in each category. You may not realize it, but you have experience in each of these categories.

For example, if you have ever worked in the retail business, you have informally trained to be a police officer. By dealing with the public, you have improved your communication skills, interest in people, problem solving skills, judgment, and a desire for self improvement. Interaction with the public on a daily basis has given you the experience of satisfied as well as unsatisfied customers. You have attempted to resolve issues and problems to complete a sale. Whatever jobs or experiences you had in the past can easily be used to boost your net worth as a law enforcement or fire candidate.

Twenty-five years ago, the prerequisites and expectations were much different than today. The average applicant had a high school diploma and some college (higher education was not emphasized as much as it is now). Obtaining your college degree is one of the most worthwhile milestones you can accomplish.

Co-Author, Steve Winston, tells his story. I was 26 years old, had a few units from a local college and had been working since I was 15 years old. I had lived on my own since I was 19 and was financially independent. When I applied, I had been married for five years and was working in the retail furniture business as the store manager. My responsibilities were buying the replacement furniture, making sure the deliveries were made and the customers were happy. I did not realize at the time that I was in training to become a police officer. To make a long story short, the big furniture stores were running all the smaller stores out of business, and the organization I was working for was to be among their victims.

A friend who worked for the Santa Ana Police Department urged me to apply. As I sat in front of the oral interview panel they asked the million dollar question, "What have you done to prepare for the position of police officer?" I had never taken an oral interview before and was not prepared for even the most obvious question. At the time, there were no study materials and definitely not a book like this one in existence. I paused for what seemed like minutes and sat in the chair doing my best to disguise the stress of silence. The panel felt my pain and tried to help by suggesting things a good candidate would have done. They asked, "Have you taken criminal justice classes?" "No sir." "Have you done any community service?" "No sir." "Been in the military?" "No sir." "Been on a ride along?" "No sir." "Well at least you have your degree, right?" "No sir."

I felt my chances slipping away and knew it was up to me to save myself. When the exasperated panel member said, "You know son, we have over 500 applicants for this job; why in the world would we pick you?" And then it hit me; all my life experience and a reasonably clean background would finally pay off. I replied with great confidence and enthusiasm, "Sir, the reason you are going to hire me is that I have something the other 500 applicants don't have and you can't give it to them." The panel glanced side to side at each other with a 'this ought to be good look' and said, "Alright, we'll bite. What is it?" I looked each panel member in the eye, leaned forward in my chair and said, "I have the ability to use this," as I clenched my fist in the air; meaning I could handle myself in a fight, "but, I'm an expert at using this," as I pointed to my brain. "I can get people to do what I want them to, and make them think it was their idea. It is a gift and I want to use it to be the best police officer in the city."

It was obvious that I had struck gold because the panel looked at each other with stunned reactions. Sixty seconds prior they were ready to send me home and had now asked me to step out of the room. A few minutes later they called me back in and advised me that I was just what they were looking for and **I was hired**. Timing is everything in law enforcement. The profession was transitioning from being physically imposing and intimidating to a community policing philosophy, and fortunately I fit the bill. I have loved my career ever since.

HOW TO MAKE YOURSELF A BETTER CANDIDATE
Preparation

To be a successful candidate, you must begin to prepare a year or more in advance. There are many ways to develop a public safety mindset, as well as prove to hiring officials that you deserve and are ready for the opportunity. Many agencies offer ride-along programs for students and community members. If you have any friends in law enforcement, they may also assist you in preparing to be an officer. When you go on the ride-along be respectful and take in as much knowledge as you can. Stay reasonably quiet and keep questions to a minimum unless the officer appears to welcome them. Remember, they not only have to worry about their own safety but yours as well.

Career-Focused College

Enroll in a college that specializes in criminal justice and offers a variety of law enforcement courses. Career-focused colleges have instructors with expertise in the fields they are teaching and often have associations with background investigators. Attend seminars regarding any subject related to public safety, affording you new perspectives on a variety of subjects as higher learning tends to do. This will give you a breadth of knowledge which may assist you in answering questions from an oral board and sound impressive in an opening statement

Community Service

Participating in community service programs is an honorable thing to do and also gives you a broader base of knowledge. For example, assisting at a homeless shelter will give you insight into dealing with the homeless when you become a police officer. Helping in an addiction program will assist you in learning the street verbiage of narcotics users, which may enhance your skills in medical aid situations as a firefighter. If you have previously performed community service, it demonstrates your desire to improve the community -- an important aspect of being in public safety.

Any job or volunteer position related to public safety will be helpful in the hiring process. The above picture is an automated fingerprint system. Most volunteer positions, especially those involving children, require a record check and fingerprints.

Internal Motivation

One thing that your school, academy, mother, father, or teachers cannot teach or give to you is motivation! You must be driven if you want to get into this career field. You must have an "I can do it" mentality, a "never say die" attitude; the type of person that when you hit a road block, you look for another way. The majority of people are impressed with effort; the people that succeed in life are usually the same people that also failed the most often, but persevered. If you want this career badly enough, you will find a way, and there is always a way!

Should You Go Into the Military?

The military is a once in a lifetime experience that teaches discipline, team play, how to follow rules and regulations, not to mention serving your country with honor. You will emerge a much more mature person for certain. The military also offers unbelievable training that is technologically a decade ahead of the civilian world.

Many oral panel members have served in the military and will obviously look favorably upon veterans. Service to our country will always bring about a sense of pride to a panel and the candidate themselves. Veterans are usually confident in interviews and present themselves very well as they have actual life experiences to relate to the board. The applicant with military experience may have an advantage over those who may have difficulty placing themselves in certain scenarios and articulating how they would handle situational questions.

All government agencies, including police and fire, offer "veteran status," which includes extra points added to your overall score for service to our country. In addition, those who served in an actual war zone will be awarded additional points on top of veteran status. A score of 92 will turn into a score of 97 and could be the difference in being hired or languishing on a list waiting to be hired. In public safety, you should be and are rewarded for military service.

Some Characteristics of Public Safety Employees

After years in the background business, it is evident that public safety applicants share common traits: Honesty, a desire to help people, a need to do what's right, typically athletic, muscular and strong, and most have several interests and hobbies.

Getting Started Job Hunting

In any career field, one of the most difficult things to do is to find the perfect place to work. Finding the department that fits you as well as you fitting them is the key. There are so many variables to consider such as location, climate, socioeconomic status and affordable housing. Moving to a different part of the country to live and work can be very difficult, but it can also be an incredible journey and a new beginning.

There are literally thousands of agencies that offer law enforcement and firefighting opportunities. The positions listed below are options in the public safety career field that you may or may not have considered. Do not rule them out because you think they are not as glamorous or desirable as your first choice. They could be the right fit for you.

Law Enforcement Positions:

Police Departments; Sheriff's Departments; Highway Patrol; Child Support Enforcement; University Police; School District Enforcement; Tribal Law Enforcement; Airports and Harbors; Corrections; Conservation Enforcement; Federal Law Enforcement; Homeland Security Agencies; Military Police; Marshal's Departments; County Coroner's Office; District Attorney Investigators.

Fire Department Positions:

Firefighter; Paramedic; Fire Inspector; Fire Marshal; Department of Forestry; Bureau of Land Management; Airport Fire Departments; Military Bases (civilian); Military Contractors; Municipal Agencies; State Agencies; and, Indian Casinos and Reservations.

The following is a breakdown of law enforcement agencies and number of sworn personnel for each state. These statistics were taken from the National Directory of Law Enforcement Administrators. If you are looking for the right place for you, may we suggest you begin your research by ordering a copy of the National Directory Book from the website www.safetysource.com for $129 (includes shipping and handling); expensive, but well worth it if it launches your career! It contains a listing for every sheriff's department and municipal police department in the country. This book will assist you in determining if the agency is the right size, location and political climate for you. It contains the number of sworn officers, addresses and phone numbers. This is an excellent resource to begin your research and obtain applications.

Police Officers:

State	Departments	# of Sworn Officers
Alabama	391	9161
Alaska	45	1143
Arizona	98	11266
Arkansas	333	5217
California	465	69687
Colorado	221	10697
Connecticut	105	6690
Delaware	38	1066
District of Columbia	1	3615
Florida	352	41024
Georgia	528	22381
Hawaii	6	2919
Idaho	116	2765
Illinois	932	36789
Indiana	504	11319
Iowa	388	4913
Kansas	338	7076
Kentucky	376	6858
Louisiana	344	21223
Maine	128	2497

Maryland	104	12039
Massachusetts	354	18101
Michigan	557	20937
Minnesota	463	8106
Mississippi	323	5779
Missouri	592	12761
Montana	114	1383
Nebraska	243	3124
Nevada	33	6543
New Hampshire	228	2386
New Jersey	502	24613
New Mexico	112	3673
New York	524	67477
North Carolina	438	20529
North Dakota	117	973
Ohio	860	28,594
Oklahoma	391	6857
Oregon	171	5618
Pennsylvania	1107	23,364
Rhode Island	43	2,343
South Carolina	237	8,227
South Dakota	157	1,084
Tennessee	340	13,456
Texas	967	55,167
Utah	118	3,889
Vermont	65	1,008
Virginia	290	18,673
Washington	236	8,757
West Virginia	226	2,394
Wisconsin	560	13,238
Wyoming	79	1,295
Total	**16,262**	**680,684**

Contact Information:
Directory Statistics
www.Safetysource.com
800-647-7579
Fax 715-345-7288

National Public Safety Information
PO Box 365
Stevens Point, WI 54481-9896

For prospective fire applicants, you can obtain a copy of the Western Division Fire Service Directory, which contains the names of fire chiefs, addresses and phone numbers of their respective department. This listing includes all the fire departments in Alaska, Oregon, California, Nevada, Idaho, Arizona, Hawaii, Washington, Montana and Utah.

Make $20 checks payable to the OFCA and request a copy of the directory:

OFCA
727 Center St. NE suite 300
Salem, OR 97301

My brother, a college softball coach, told me that if a person is a good enough athlete and wants to play bad enough, there is always a place for them. Public safety is the same way. If you are good enough, and you want this career bad enough to make the necessary sacrifices, there is a position for you. Your future employment depends on your determination and research. Start with finding "your department."

Reputation of the Department

All departments are not created equal. There are many factors to consider when deciding with which agency to apply, the first being their reputation. You want to be proud of where you work. Check with sworn public safety officers from the area and determine if the agency has a good reputation. Is it on its way up or down? As you have seen on television, one well-publicized incident can taint decades of good deeds by a police or fire department. A bad reputation translates into less of a partnership with the community, increased tension on the streets and many more assaults on officers and firefighters. It will also affect your pocketbook, as less community support results in smaller raises and sometimes even pay cuts.

SWAT High-Risk-Entry

Each respective department has its own unique strengths and weaknesses. Becoming a SWAT Member may be a long-term goal. Do your homework and be able to give reasons why it's a career objective.

Major Narcotics Bust

Do you have an interest in narcotics? It will take you many years to achieve the goal of selection to a major narcotic unit, but if you bring up an interest in an interview be sure to have a basic understanding of that particular unit. Once again, do your homework.

Does the city have a gang problem? How much esteem does this particular gang unit carry throughout the county and state? Why do you want to be in this unit, with this department? Perhaps you have grown up in a gang area and have observed the intimidation and oppression of law abiding citizens firsthand. This would be an excellent point to bring up at some point in the interview.

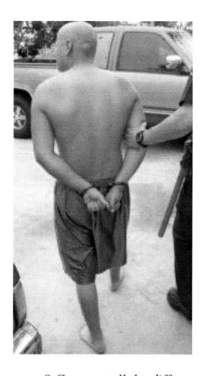

How street-wise are you? Can you tell the difference between a gang member and an individual who simply shaves his head? What is the definition of a street gang? Society will depend on you to protect them from gang members.

Philosophy of the Department

Do your research

It is imperative that you select an agency that has a similar philosophy as your own. If you are a hard core conservative republican and the department has a liberal viewpoint (such as focusing a high percentage of resources on community outreach programs as well as other democratic agendas), philosophically, you may have a difficult time believing in and adjusting to their methods. Conversely, if you are a democrat and lean to the liberal side, an aggressive enforcement type departmental atmosphere may not work for you. It is important to get a feel of the managerial philosophy before you commit to a career in which you are a square peg in a round hole. This information can be difficult to obtain, though it is possible. Call the department and ask to speak with the station supervisor; level with him or her and they may give you some good information. A caveat is not to take everything they say as gospel; simply use it as another corroborating source of information.

Cost of Living

Determine if you can afford to live near the department with which you are applying or if you would need to commute a significant distance to work. There are officers in some parts of the country making well over $100,000 a year who still need to drive an hour-and-a-half to get to work. This may seem like an easy compromise at the beginning of your career, but it will wear on you and add stress as your commute time increases with the ever increasing population.

Diversity

Are you comfortable with the racial makeup of the community? Do they speak languages you are not familiar with? What is your life philosophy and are you willing to compromise it?

Do you want to be part of a large department, or a smaller department with job requirements not typical at larger agencies, such as barking dog calls or courtesy vacation checks on a citizen's home? Smaller agencies have different aspects to consider – lifestyle, standard of living, and the availability of special detail positions within the department. The employees will usually maintain more of a family type atmosphere than a larger department, which can be positive or negative depending on your personality. The department and community will probably have closer ties, but be restricted with limited funding. Fiscal issues make it more difficult to implement progressive programs requiring new technology. Depending on the needs of the community, there may or may not be a need for state-of-the-art programs and one must consider what type of law enforcement they wish to be involved in. A smaller agency has very little turnover or attrition, limiting promotional opportunities they would have at a larger agency. Consider your long-term goals and select a department that will match your needs.

Each type of agency has their own unique pros and cons, so it depends on you and what your wants and needs are. Hiring standards will vary greatly between agencies and you should research them prior to applying to determine if you meet their standards. A smaller agency is often neglected by potential applicants due to a lack of research and the limited exposure of small towns. They also have less revenue to advertise and are most likely relying on word of mouth to attract potential employees.

One of the gauges we suggest is to grade yourself (A, B, C or D) as a potential candidate for an agency. Then give the potential agency a grade based on your research. An 'A' agency usually considers 'A' applicants or 'B" applicants who have the ability and desire to become an 'A' applicant. 'C' agencies look for 'C' or above applicants, etc. To be a serious candidate for an agency, make sure you and the agency you have selected are a good match and be realistic. Oftentimes, applicants apply at two agencies -- one a very busy city and the other a very quiet city. An individual with a hyperactive personality would match the busy city and others may be more satisfied with a slower activity rate. In an oral interview, it would be hard to justify the two applications and it could be perceived that you have not done enough research or you just want a job. Remember, in any case, **"The prepared applicant is the one who gets hired."**

Race and Public Safety

Demographics of the city are something to consider when choosing a department. Chances are you will be in a shooting or use some type of major force causing serious injury or death at some point in your career. No matter your race, if it involves someone of another race, there will be additional scrutiny and possibly an independent investigation. Below are two examples of racially motivated media attacks on police officers:

A few years ago officers working for a large agency responded to a call of a suspicious vehicle in a parking lot. The officer approached the locked vehicle and saw a young female foaming at the mouth and

64

holding a handgun in her lap. They developed a plan (not the best plan, but well intentioned) and broke the driver's side window and attempted to grab the gun. When they broke the window, the female was startled and reached for the gun. The officers opened fire and she unfortunately died. Three officers and a sergeant were fired and a multi-million dollar payout was made to her family. Political pressure caused the federal government to take over the department. Very little was ever mentioned about the gun in her lap or the stolen shotgun in the trunk.

A few weeks earlier in a neighboring city, a young male led police on a high speed pursuit. He finally pulled into a driveway and reached into his waistband simulating a weapon, prompting the officers to shoot him. He died at the scene and not one community activist made an issue of the incident. Thankfully, the officers were not injured and their use of force was justified; however, the question should be asked why community activists allege racism in a case where the deceased was armed with a gun and not in the other case when the individual was unarmed?

How does this affect you going into public safety you may be asking? It is the environment created by the media and society in general, and it will alter the way you do your job. It is difficult enough to make instantaneous decisions without injecting race into the formula. Doing the right thing becomes convoluted. For example, most people fear being called a racist even by a person who is saying it for their own agenda; usually by a person who is trying to avoid a ticket or an arrest by intimidation. Our take on this subject is that any change from what your original intentions were in the first place due to race is wrong. It is just as wrong to let someone go because you don't want to "deal with it" as it is to arrest or cite someone because of their race. If you are intimidated by a specific race of people or will succumb to threats of filing a complaint against you, you may want to reconsider working in public safety. Making the right decision is based on the sequence of events and not on race.

You need to be comfortable where you work and with the demographics of the community. As this country becomes more urbanized and apartment buildings become increasingly dense with immigrants, there will be more political situations when it comes to use of force and deployment of fire personnel. Imagine making the difficult decision not to go into a burning apartment building because you know you will be sending the firefighters to their deaths. The reason according to the media and activists will be because of their race. You will be defending your actions to people, including the court system, who have their own agenda. Conduct a self-analysis of your experiences and determine if you can be fair and objective regardless of any prejudices you may have.

Tax Benefits
Taxes hurt all of us and it is in your best interest to pay as little as legally possible. Save every receipt that is involved in your attempts to obtain a public safety position, such as restaurant, gas, travel, clothing, and educational receipts. Occasionally, you will fly to your appointments and stay in a hotel; ensure you get a business card from that agency. You will also need to develop resumes, utilize postage, stationary and possibly hire a service to do this for you. You may belong to several trade magazines, which you'll need in order to keep abreast of the current information in your career field. Maintain a separate file for your job-seeking expenses and be as organized as possible.

Specific Agency Qualifications
Once you have decided public safety is right for you, you will need to determine whether or not you meet the minimum qualifications. It is important to realize every agency is slightly different regarding what is acceptable and what is not. Below are general requirements, but make sure you speak with a background investigator for specifics at each particular agency. This includes both published and unpublished criteria to determine if you meet the minimum qualifications. Be honest with them and understand it is in everyone's best interest to advise you early in the testing process if you are wasting your time.

Here is an excellent example of a quality applicant and a quality department failing to collaborate. A good friend of mine applied at a local police department not knowing he could not be considered for employment due to its policy on pending litigation. An "extremist group" filed a frivolous lawsuit against him and his coworkers. The lawsuit was ridiculous on its premise, but it didn't matter due to policy. The policy at that agency has since changed, but it was too late for him. Each agency has different policies and standards, and what may be important to one department may be insignificant to another. So level with the background investigator before you apply if you have any concerns. Some of the investigators will be upfront with you, and others due to department policies would not be willing to discuss the matter until you are in their process. Some background issues are covered later in the book and may be very helpful to you making an informed decision.

Background investigators were appointed to the position due to a number of qualities they possess, such as fairness, objectivity, honesty, integrity, and communication skills. They are representatives of the chief and want what is best for the department; they also want what is best for you. Some background investigators will take the time and effort to offer constructive feedback and tips that will give you the tools and information to rebuild yourself if you do not qualify. They may also direct you to other agencies that may not disqualify you for what occurred. Dozens of applicants who did not qualify for our department for one reason or another are now productive police officers at smaller agencies. If you make a good enough impression, the background investigator may even place a call for you. You may also want to consider a three or four person department in a remote area. They often need officers and do not have a large recruiting budget, so they only get a few applicants.

Listed below are the general standards used for police officers across the nation. These minimum standards are rapidly being adopted by all police and fire agencies. Whether or not you will be held to these standards, compliance would be a good place to start in getting yourself hired.

Minimum Standards for Employment in Public Safety
Every Police Officer employed by a department shall be selected in conformance with the following requirements:

- **Felony Convictions**: Government Code Section 1029 limits employment of convicted felons.

- **Fingerprinting and Record Check**: Government Code Section 1030 and 1031 requires fingerprinting and search of state and national files to reveal any criminal records.

- **Age**: Government Code Section 1031 requires a minimum age of 18 years; however, a particular state can raise this requirement.

- **Moral Character**: Government Code Section 1031 requires good moral character as determined by a thorough background investigation.

- **Education**: Government Code Section 1031 requires a high school diploma, passage of the General Education Development Test (GED), or attainment of a two-year or four-year degree from an accredited college or university.

- **Medical and Psychological Suitability**: Requires an examination of physical, emotional and mental conditions.

- **Citizenship**: Either a current citizen or a resident alien who has applied for citizenship.

- **Must Be Personally Interviewed:** Prior to appointment by either the department head or a representative to determine the person's suitability for law enforcement or fire service. This includes, but is not limited to, a person's appearance, personality, maturity, temperament, background, as well as the ability to communicate and interact with others. The same regulations, in turn, must be satisfied by the department representative participating on the oral interview panel.

- **Must have reading and writing ability** and be able to verbally articulate him or herself and write at the levels necessary to perform the job of a peace officer or firefighter as determined by the criteria of the department.

- All requirements also shall apply to each lateral (transfer from another agency) applicant regardless of the rank in which the person was appointed at his current department.

Education

How Much Do You Really Need?
Do society and yourself a favor and go to school!

Most agencies will state that you can apply for a public safety position with at the very minimum a high school diploma or a general education diploma (GED). Very few agencies require an Associate's Degree and even fewer require a Bachelor's Degree. Now let's be serious. Do you really think you will be hired with just a GED? A limited education limits your opportunities as well as your personal growth! Now that you have been told the cold hard truth, get started accumulating college units. You should be immersing yourself into criminal justice, law and government classes. Do not focus just on personal interests; understand the program degree requirements and speak with a counselor to obtain your degree in the most expeditious manner. I know people with enough college credits for two Bachelor's Degrees, but they do not even possess one!

Consider a career college which specifically focuses on criminal justice; they hire current and retired professionals from the field who bring real world expertise to the classroom. Attendance at these schools displays your motivation and drive to succeed. These types of schools and programs offer a personal touch and assist you in achieving your goals.

A degree is something that cannot be taken away from you and will impress a background investigator, who most likely began his career before a college degree was considered necessary or even important. Remember; do not be pompous or arrogant if you do possess a degree, as these days it is becoming more common to have one than not. It has been said that today's Bachelor's Degree is equivalent to yesteryear's high school diploma. Higher degrees such as Master's and a Ph.D. are also not uncommon. That being said, possessing a degree or not possessing one will be an important factor, but not the deciding factor in whether or not you get hired. It is imperative you understand that there is a correlation between your level of education and success on an interview panel. Those with a higher level of education tend to be more articulate, thus increasing their chances for success. What the department wants to see is a concerted and demonstrated effort to prove you have been working toward public safety for some time and the ability to set a goal and attain it.

The Application

Number of Applications You Should Submit – <u>Law Enforcement</u>
Applicants occasionally submit applications with numerous agencies (four or more). These applicants were probably given bad advice by well-meaning mentors. Understand that on the law enforcement side, too many applications make you look desperate and willing to work for any department! Even if you are desperate and willing to work anywhere, do your best not to give that impression.

There are a few ways the fire and police processes are different and the number of agencies you apply for is one of them. This chapter is specifically for prospective police officers only and the next chapter for prospective firefighters only. Choose your agencies carefully and know why you selected each one. Show consistency. For example: If agency X asks why you applied there and you say it is because you want to work at a large agency, yet you applied at three smaller agencies, your credibility will be called into question.

Agencies are going to ask you, "Why do you want to work for us compared to the other twenty agencies you applied with?" After many years of hiring people we have yet to hear a good answer to this question. Choose two or three departments you would love to work for and have a reason you selected each of those agencies, such as Department X has a strong SWAT Unit that is highly respected throughout the state and I would like to be part of that program someday; or, Department Y has a world renowned Search and Rescue Team that I would be honored to participate in; and, Department Z is internationally known for its community outreach program.

Number of Applications You Should Submit - <u>Firefighter</u>
Firefighter positions take more groundwork and effort into the application process than law enforcement does. A prospective firefighter should apply everywhere and be willing to move anywhere to obtain a position. Almost all firefighter resumes look the same, so you will need to standout in the oral interview. We will cover that later, and trust me you will stand out! But for now, apply everywhere you would like to work and be prepared to travel for the testing. If you're not prepared to travel or move, what's the point in applying at that particular agency? Agencies across the nation have a standard 30% no-show rate for various examinations. Amazing! Why would an applicant take the time and make the effort to complete the application if he was not going to make an effort to show up for the test? Many agencies will label you as a "no show" and will not allow you to re-enter the process for a specific period of time (usually six months). Also, other agencies might inquire about

your status and it will be reported as a no show. They may deem you as irresponsible and unreliable. If there is an emergency or a situation beyond your control, simply call the agency and explain the situation. In most cases they are willing to reschedule your exam. So, if you applied for a position, show up to take the test!

An application paints a picture of you with words.
Good Application and Bad Application

Consider the application just another opportunity to place your self above the rest of the candidates. Consider it a test that you should take your time with and follow all of the instructions. An outstanding application will not get you hired, but an incomplete or sloppy application will eliminate you immediately.

Perception is a powerful concern in the field of law enforcement and firefighting. Whether they are right or wrong a person's perception is reality. You are being evaluated from the time you pick up the application until the day you are hired and then throughout your career.

Submitting the Application

The application speaks volumes in a page. Three ways of obtaining an application are by downloading it online, contacting the Personnel Division by telephone, or in person. The application is the first impression you make to police or fire personnel evaluators. So do yourself a favor and take this opportunity to separate yourself from the rest by dressing in a business suit if going in person, or by being respectful and professional in your conversations and e-mails; you never get a second chance to make a first impression.

Sloppy handwriting will create the opinion you are unprofessional and unorganized. Ensure everything you write is neat and spelled correctly. If necessary, visit a police and fire uniform store and purchase a "writing template," which will assist in writing legibly, but if possible type it. I also highly recommend that you have someone proofread any correspondence, including your application and resume. Simple mistakes will state you are either not serious and/or an unqualified amateur.

Ensure the dates and employment history is correct by calling each employer and verifying the dates as well as contacting your supervisors and co-workers if they are still employed there. This will help ensure they remember you and will consider it a sign of respect that you called to ask them to be an employment reference.

Simple errors make you appear less than professional and possibly untruthful. One of the most common mistakes on an application is failing to sign your name at the bottom. Omitting pertinent information shows inattention to detail and may call into question your overall competence. This will also annoy the personnel representative as she must re-contact dozens of applicants and have them sign their applications to provide the missing information; creating the mindset that you do not belong in the process. Some agencies might not be this considerate and will disqualify you for submitting an incomplete application or one of poor quality

Always make a copy of the paperwork you submit
Do You Need a Resume?

If you want to separate yourself from other applicants, attach a professional résumé on high quality paper to the application. Make sure the résumé is perfect by having it proofread by several people. Even if you read it a dozen times you may not catch every error that other people will identify immediately. Also, you could conduct online research or reference a grammar book for the exact way to properly format a résumé. If you are a young applicant, your résumé may not be very long. There are differing opinions on the proper etiquette of a résumé and the number of pages it should be. Some grammar books state a résumé should not be more than one page and other books say two to three pages is acceptable. We suggest you keep the résumé to one page in 12 font Ariel style. This is succinct and powerful.

Furthermore, most personnel representatives and oral interview panel members agree that one page is not only sufficient, but preferred. They are pressed for time and will only <u>glance</u> at your resume focusing on key words and dates. Just as you did to the application, ensure that the dates are correct and the particulars do not contradict the information on the application. Only approximately 10-20% of the applicants will attach résumés, which will immediately separate you from the other candidates, and only about half of the resumes attached will look professional, further elevating you into the elite. Before you even step into the interview room, **you will already be in the top ten percent.**

Wait for Testing Date Notification

After the application is turned in, you will have to wait to test until a group of applicants with large enough numbers is formed. The testing is usually within a couple of months of your application submission date.

Weight

If you find yourself close to the physical agility testing (PAT) and are not in top physical condition and need to drop a significant amount of weight, you will need to go immediately on a strict weight-loss program. Your future career depends on your being in great physical condition. You need to get extreme and you need to do it now. Medical experts will tell you that sudden and significant weight loss harms your body. We believe it is more harmful to be overweight and out of shape than making an effort to get lean. What will harm you even more is not obtaining your dream because of your weight, so make the decision right now to be a warrior and get fit!

75

What you absolutely do not want to do is create a bad first impression and imply to the PAT evaluators that you have a weight problem and could not handle a public safety position. They will wonder what your out of shape body will look like in fifteen years. You must have the ability to chase suspects, jump over walls and carry people out of burning buildings. Being an overweight superstar does not matter if their perception of you is tarnished by your appearance.

There are a lot of healthy diets that are very popular right now; consider trying one to help get you on track. Remember, you never get a second chance to make a first impression, and if it means sacrificing in the short term for a long-term goal it will be well worth the effort.

Before you do any type of rigorous dieting, get a physical with a doctor to rule out diabetes or any other underlying health problems. I would normally advise you to eat balanced meals and exercise regularly, but if you're currently significantly overweight and the PAT is approaching, it's too late for that. You should consider going carbohydrate free until the physical agility test day. You can drop approximately twenty pounds in one month and you will feel great.

These diets are extremely simple; just cut out sugars, bread and rice. I guarantee you will begin to get compliments within a couple of weeks. Your confidence will shoot through the roof and it will also lower your agility times and assist you in achieving your ultimate goal of getting hired.

Physical Agility Testing (PAT)

Depending on the agency, the PAT or the written test may come first. To excel in the physical agility test you must first learn what the test is about ahead of time and train for each particular component. For example, at our agency we have created a series of components based on what we believe simulate actual law enforcement situations.

There are different types of PAT's with different names, such as physical abilities, physical assessment, etc. Some of these PATs consist of events you may believe have no connection with the position. There is most likely a reason for that test based on the history and tradition of the department.

Every agency determines what is important to them. For example, one particular department has the recruits run a one-and-a-half mile course. Other departments shorten the course but expect faster times. The PAT must comply with a federal law requiring the test be restricted to "essential job functions". If a specific requirement is important to that agency, it should be important to you if you want to work for them. You need to train to excel in that particular requirement.

You should give an all-out effort during the physical agility testing; less than a 100% effort will make a bad impression. Act like your life, your partner's life or a family member's life depends on the component you're competing in. If you are injured while completing the PAT and are still able to finish, do so. Obviously, if you are seriously injured, stop and seek medical attention, but anything that will not result in long-term damage should be brushed aside both mentally and physically.

You must treat the physical agility test as if it were a real situation in the field. Are you going to leave your partner inside of a structure fire and let him die alone? Will you physically and mentally give up during a fight with a murder suspect because you hurt your shoulder? Prove to us that we can count on you in a real situation, and at this point the PAT is the only way to prove that you are physically capable of performing the job. The PAT shows what type of person you truly are and how much heart you have when it really matters.

One applicant was surprised at the difficulty of the PAT and was visibly exhausted after the first component. When we called his name for the second component he was nowhere to be found; we located him under a tree, sitting cross-legged, palms up, eyes closed in deep meditation. It didn't help his score and he does not work for us. He was not prepared physically or mentally to give a 100% effort. Do not ask the evaluator if you passed or how you did as it demonstrates a lack of confidence in yourself; you should know you passed!

Agility testing measures your ability to perform essential job functions. The actual test usually mimics what occurs in the field. For example, a police department test often consists of four components, which are outstanding indicators of your ability to perform at the necessary level of a police officer. Below is a breakdown of each component:

- First component is going over a six-foot wall while maintaining control.
- Second component is a simulated wounded officer (165 lbs.) whom you must carry or drag 25 feet.
- Third component is a short pursuit, which requires quick agile movements as it includes going over a four-foot wall and numerous turning and zig-zag movements to avoid obstacles.
- Fourth component is a stair climb with approximately 25 steps testing your ability to use bursts of speed combined with agility.

It is crucial to impress the officers and firefighters that are watching. Trust us when we say, they will remember you if you if you give a substandard effort. Let's assume you are a physical superstar and you can easily pass these tests with a less than 100% effort; do you think an officer or firefighter wants to work with someone of that mentality? Even if you are able to easily complete the components, do not coast as it will cost you in more ways than just your score! The evaluators are trained observers and experts in human behavior as their profession demands; they will know if you are not giving your best effort, so do not risk it. Also, to further impress your evaluators cheer-on and encourage the other candidates to do their best.

Most fire departments will be comprised of the following physical agility tests:

- Putting on a turnout jacket, helmet and gloves with a self-contained breathing apparatus.
- Jump a 6-foot wall then advance a charged 1 ¾" hose line 100 feet.
- Pound a sledge hammer over your head 30 times (simulating using an axe on a rooftop).
- Drag a dummy 20 yards and then carry an oxygen bag and first aid kit 50 yards.
- Raise a halyard on a 35-foot ladder (all the way to the top then all the way down two times).
- Climb a ladder to the 3rd story, lean over the railing and raise a 3-inch hose bundle up to the window sill then back down. Climb down the ladder, and when your feet are safely on the ground, the time stops.
- This entire test is all inclusive with no rest. All fire departments will have some semblance of these types of components. These physical ability tests are generally "Biddle" certified due to validity checks and relating specifically to job performance.

Preparation for the PAT
Upper Body Strength

If you are a television viewer, start by doing push-ups during commercial breaks. Start at 25 per commercial break and work your way up to as many as you can do. You will be surprised how quickly you improve in this area. In a month, you'll be up to 100 at a time. Females will have to work harder due to a gender-based lack of upper body strength. Ladies, the academy doesn't care that you have a genetically weaker chest, just as a parolee won't care in the streets when he is trying to take your gun. It is a fact that you will have to work harder than your male counterpart to meet the standards in this area.

Cardio Vascular Fitness

The physical agility test is more than a test of your physical fitness; it is a test of your heart and measures your determination to be a successful police officer or firefighter. In both professions, you will be physically and mentally tested beyond the capacities of the average person. In public safety we depend on our partners to give their all and assist in our success. If just one person gives up, it weakens the team and others may be hurt and/or killed.

As background investigators, we watch and evaluate effort and heart. Recently, a young officer from a local agency tested and failed the physical agility. Worse was my observation that the young man who looked extremely athletic was not giving a full effort and thought he would breeze through the test. He does not work for our agency and with that attitude will have a hard time being successful anywhere. In another case, a young man could not make it over the six- foot wall. I took him aside and he told me he just couldn't do it. I explained with that attitude he was right. I said, "I'm more than twice your age," and then quickly climbed over the wall. "You have to want it bad enough or you are not mentally ready for the job." The pep talk worked. He spent a few seconds of internal motivation and then went over the wall, which had beaten him just moments before.

Your cardio conditioning should comprise of running, which includes interval training; that is, to sprint occasionally during your run. Work your way up to three miles at a good pace and at least once a week spend the workout running 100-yard dashes. There are numerous other ways to get fit, and hiking is one of them, especially for firefighter candidates. Carry a heavy backpack and walk steep terrains and you will be ahead of the game on the stair climbs. Any type of martial arts or bicycling will also be beneficial.

Attire

The way you dress creates an immediate impression of you. Do not, and I repeat do not, wear shirts with any type of bold statements, hip-hop, rap, rock 'n roll, or anything that may be considered controversial or unprofessional. Ladies, be sure and wear proper support such as a good sports bra(s) to aid yourself in the running events. Also, and this is for everyone, do not wear baggy clothing as it will not only slow you down, but look unprofessional.

Regarding footwear, we've given this advice to hundreds of applicants and only one took it. Coincidentally, he is now working for our department. Bring along cleats for turf components as they will improve your time, especially when the grass is wet. Since you have read this book, you would have conducted research prior to the actual test and already know what the agility test consisted of and whether it requires running shoes, cleats, knee pads, etc. Besides having a better score on the agility test, something like this indicates many other attributes of the candidate such as logic, maturity, intelligence, preparation and a willingness to think outside the box. It will also impress the background investigator or physical agility evaluator.

Civil Service Test

Most local public safety departments will conduct an initial screening tool by administering a general knowledge test, also known as the civil service test. There are many ways to study for this test, but the quickest and possibly most effective method is to visit your local library and/or utilize our written test guide. The library will have several different types of civil service tests available for your viewing. Doing this will no doubt assist you in passing the examination as there will be the actual questions from older exams; these questions will give you an idea of what you will encounter. There will be very few surprises if you research in advance which usually minimizes angst.

Most public safety hiring officials recommend applicants visit the Police Officer Standards and Training (POST) website as a research avenue. The POST website, www.post.ca.gov, offers a sample practice test at no cost to the applicant. If you take the test and fail it several times, you should consider enrolling in additional English and composition courses in college.

We offer a Written Test Workbook online, but any type of practice test will be beneficial in bringing you into the test-taking mode. There are many great books on this subject at your local library and bookstore. Spend the money for the book! People will spend thousands of dollars to travel to a test, but balk at spending $25 for a book that will assist them in getting hired. Make an investment in yourself.

Below are sample questions from the California Commission on Peace Officer Standards and Training:

1. Every place you see a dashed blank line, you supply the missing word; for example, a sentence might read, "The driver was injured when his _ _ _ crashed into the tree." The word 'car' or 'bus' could have been used, but the words 'truck', 'auto' or 'vehicle' would be incorrect.

2. The juvenile justice system encounters many types of people. Here is one example. Mary was 15 when _ _ _ was first arrested. Mary was a popular girl. _ _ the time of her arrest, _ _ _ was doing well in school. According to _ _ _ teacher, she was a hard working _ _ _ _ _ _ _ and made excellent grades. Mary also had _ good attitude toward school. Mary lived with her parents _ _ a middle class neighborhood.

3. Choose the Correct Sentence:

a. Jail is not a pleasant place to be, but they do get their basic necessities met.
b. Jail is not a pleasant place to be, but prisoners get their basic needs met. (correct)

a. The suspect disliked the officer as did the judge. (correct)
b. The suspect disliked the officer as much as the judge.

a. The officer was hoping to get a new partner, one that had a great deal of experience. (correct)
b. The officer was hoping to get a new partner. One that had a great deal of experience.

a. Bullet fragments were gathered by officers in envelopes.
b. Bullet fragments were gathered in envelopes by officers. (correct)

a. In rural areas, more deer are killed by automobiles than by hunters. (correct)
b. In rural areas, automobiles kill more deer than hunters.

a. The next time Mary was in town, she agreed to have lunch with Sue. (correct)
b. Mary agreed to have lunch with Sue the next time she was in town.

4. Mark the Correct Answer:

The witness <u>corroborated</u> the suspect's story.
    a. verified   b. contradicted   c. added to   d. questioned

It was not a very <u>pragmatic</u> plan.
    a. plausible   b. serious   c. practical   d. sensible

Police <u>sequestered</u> the suspect.
    a. caught   b. isolated   c. arrested   d. released

In order to conceal her guilt, Linda told a <u>blatant</u> lie.
    a. harmless   b. subtle   c. careless   d. obvious

The best and fastest way to improve vocabulary, in addition to comprehension and writing ability is to read, read and read some more. It is only through using and interacting with language that true mastery achieved.

Choose the correctly spelled words.
    a. analisys   b. annalysis   c. analysis   d. anallysys

    a. apparent   b. apperant   c. aparent   d. aperant

    a. conscientious   b. consceintious   c. conscienteous

    a. receive   b. recieve   c. receve   d. recive

With the advent of text messaging, the ability of young Americans to spell correctly has eroded rapidly. Attempt to text the entire word and look up the words rather than guess the way they are spelled.

There will be a section on reading comprehension. The best way to prepare for this component is to read books, short stories, newspapers, magazine articles and anything else you can get your hands on. There are three basic test strategies to master this section.

1.   Read the question and responses before reading the passage.

2.   Don't answer the question based on your opinions or knowledge; answer the question based on what is in the passage.

3.   Read the question very carefully.

Notification and Making the Interview Appointment

After the PAT, you will be notified in one of two ways whether or not you passed. The first way may be via mail, in which there will be instructions on how to proceed if your preparation paid off. Follow these instructions carefully. For example, if the notification states for you to call between the hours of 8:00 a.m. and 9:00 a.m. on Tuesday, don't return the call at 9:15 on Thursday!

When you do call within the correct times, be professional; in fact, write out what you want to say on paper and tape it near the phone. This will prevent any verbal fumbling or a gust of wind from blowing the paper off the desk, and it will be in front of you just in case you become tongue-tied or draw a blank. The note should contain bullet points with a word or two on each subject you want to discuss as it will remind you what you wanted to ask or say. Remember, you may be speaking with someone from the oral panel, so do it right. At the conclusion of the conversation, repeat the day and time of the interview and give a professional thank you.

The second way you may be notified is by phone. They might call you and leave a message. This can be problematic as others may have access to your answering machine and your home phone. Frequently check old messages and ask your roommates and/or family members if they received a call from the police department. If you feel you may have missed the call and a substantial amount of time has passed (several weeks), call personnel and check on your status. Having peace of mind is much better than stressing for weeks.

Testing

Testing is most often conducted by representatives from the city and/or department. Do not consider the test separate from an interview; at all times act as if you are being watched, because you are! Your background investigator may be present and observing, and even if he is not, he will certainly be advised if your behavior is inappropriate, immature, rude, or condescending. Remember, from this point on, everything you do and say is considered part of the interview process, especially when you are in the public eye. People usually act different when they think no one is watching, but remember someone is always watching!

The Day of the Interview

If you have prepared properly you will sleep like a baby; if you haven't prepared, good luck on your sleepless night. If you have an early appointment time, wake up significantly ahead of time and eat a light breakfast. If your appointment is in the afternoon, get an hour of extra sleep; not because you need it, but because it is one less hour to stress over the interview. I also recommend having a great workout about three hours before your interview to help release tension. Listen to your opening and closing statements on audiotape and videotape one more time. Watch our DVD one more time and know that you are totally familiar with the content. Be confident as this is the day you launch your career!

The Arrival

Interviews are really nothing to be intimidated by if you are properly prepared. The first thing you must do is be on time; remember, it's better to be one hour early than one minute late. If you are in an unfamiliar area you should leave early. It would be wise to drive the route the night before the interview, but be sure to account for traffic during the day. You should arrive in the parking lot approximately 45 minutes to an hour early, as that will give you built-in extra time in case you hit a major accident or construction on the way. If you are early, you can sit in the reception area and soak up the atmosphere and culture of the department. How do they act, dress, talk, etc.? This will give you an edge and insight into the department's "personality."

Don't Be Late!

Being on hundreds of interview panels we can tell you for a fact that nothing bothers a panel more than an applicant being a few minutes late. You may think that is a little bit too strict, but imagine being on an interview panel for several days; you're tired of listening to the same speech and same answers over and over again. Believe me when I tell you, it doesn't take much to become irritated at this point. Tardiness is the epitome of being self-centered and irresponsible and will also affect other applicants as it pushes all of the other appointments back. If the unlikely situation occurs that you will be late, you must call at least 30 minutes in advance and notify the personnel representative of what occurred; this will slightly mitigate the fact that you are late, but why take a chance on any of this? Take our advice and show up substantially early and do not allow being late to be a factor in whether you are successful or not.

The Parking Lot

Think about your oral while sitting in the car, recite your introduction and be natural. When the appointment time is approximately 30 minutes away, make your way into the building. Imagine you have a camera focused on you, as a panel member could be looking into the parking lot from the office window at that very moment. Their vehicle could be parked next to yours. It will be obvious why you are there and you will be watched. Take your time walking into the building and relax; there is no rush because you left early, right? Use the restroom, wash your hands, look at your suit and tie for drips of breakfast or coffee stains. Check in with a receptionist approximately 20 minutes before your appointment and sit back and calmly watch the rushed applicants.

Oral Interview Entry

Upon being summoned by the secretary for your appointment, walk in confidently. Be careful not to slam the door, but don't shut the door so slowly, that it appears you lack confidence in yourself. This is going to sound comical, but practice this just as you would anything else.

Greeting

Greet the person nearest you with a solid, firm handshake reaching the back of the palm. Proceed to the next closest person also giving them firm handshake. Ensure that you look directly into the eyes of each person, which exudes confidence, and give a personal greeting such as "nice to meet you," "good morning," "pleasure to meet you" or any greeting you prefer. You may have sweaty palms due to nervousness. Don't worry too much about this because it is very common and panel members are used to it; however, if you are aware you have this problem and you know it's like shaking hands with a bucket of water, carry a handkerchief and wipe your hands as you are walking toward the interview room and put it in your pocket prior to entering. Practice doing this so you don't fumble when the real interview comes.

You may break out in hives, or turn bright red, but don't worry about these types of things; they are very common. We know you are nervous and how important the interview is to you.

Taking Advice From Others

We consistently hear the advice applicants received from well-meaning employees; almost all of it is wrong! We have also received information that many colleges are teaching improper interview techniques in their classes. I believe in education, but you have to remember public safety is different from the civilian world and most professors do not understand the culture involved. So do what you need to do to get an "A" in the class, but pay attention to this book!

Some people love to give advice; keep in mind that many of them are full of bad advice and something else as well. You should be very selective in whom you choose as your mentor and filter their advice. They may have begun their career before IQ mattered, or be a disgruntled employee. They were probably hired before a polygraph and psychological examination were ever part of the hiring process.

Read this book closely and utilize the tips from <u>public safety officials</u> who are **actually** on interview panels. True experts from the field you dream of entering are giving you factual, real-world advice, helping you achieve your goal!

More Do's and Do Not's
According to a hiring official of one the largest agencies in the United States, one who has conducted hundreds of oral interview panels, there are a few things you can do to improve your odds for a successful interview:

- **Be confident** - From the moment you walk into the building, a lack of confidence is a recipe for disaster.
- **Maintain a Professional and Friendly Demeanor** - This exudes more authority than a badge ever will.
- **Be Friendly -** Being likeable is a key to almost everything successful in life.
- **Display Loyalty -** Do not bad-mouth prior jobs or coworkers.
- **Be Honest -** Admit mistakes and how you improved since then. Say what you believe, but do not say what you think the panel wants to hear as you may be wrong.

There are also a few things you should absolutely **never do** during an interview:

Do Not:
- Chew gum
- Leave your cell phone on
- Turn your back to an interviewer, even slightly
- Ask questions at the end of the interview, unless they pertain to the selection process
- Interrupt panel members
- Use fillers repeatedly, such as: um, like, ok, ya know, yeah

One applicant actually answered his cell phone during the interview. Hopefully it was from a prospective employer as he will not be working here!

The Oral Interview Panel

The panel is usually comprised of sworn personnel who range from 35 to 50 years old. They are generally high achievers that have developed a good reputation with the personnel section. On occasion, they will play roles; one will be the nice guy and one will play the bad guy (good cop, bad cop). Many times they do not even know they are doing this as it almost always strikes a balance during the interview. The way you deal with a difficult panel member is to do absolutely nothing; do not change the way you have been training. Give him/her the same amount of attention and respect you are giving the others. Remember, it isn't only your answers that are being tested, it's also your personality and how you handle adversity. You also want to be yourself during the interview. Do not conform to what you think they may want; if you were not being yourself, would you really want to be hired like that? And you could be wrong!

Right or Left Hand?

As you sit down you may be so nervous your mouth goes dry or you feel compelled to cough or clear your throat due to being uncomfortable. Do us all a favor and use your left hand to conduct your sanitary business as you use your right to shake hands. When you use your right hand to cover your mouth or wiping your mouth, it is difficult for us to think of anything else. To understand exactly what I'm talking about, rent the movie "Philadelphia." There is a powerful scene in which an AIDS victim, played by Tom Hanks, is at an attorney's office (Denzel Washington). The camera focuses on each object Hank's character touches and expertly shows Denzel's nervousness and uneasiness.

Hand Movement

Do not be worried about using your hands when speaking; just be sure the movement appears natural.

Introduction and Opening Statement

It is proper to wait for the formality of being asked to have a seat before you sit down; and when you sit down, keep your feet flat on the floor and act natural without slouching!

The first thing you will be asked by the oral panel is to talk about yourself. Consider this your opportunity to give what is called a hero statement. Do not cut yourself short on this opportunity. Explain how you have prepared for this position. Don't be a "robot," let your true personality shine through. It should sound something like this:

"Hello, my name is Jonathan Smith. Two years ago I graduated from XYZ University with a Bachelor's Degree in mathematics. I became interested in the fire safety service when I was eight years old and saw the movie Back Draft. From that point on I studied everything I could get my hands on regarding firefighting. Immediately after high school I enrolled at XYZ Community College in the Fire Academy and finished near the top of my class and was awarded "Most Likely to Become a Firefighter" by the academy staff.

I knew it would be difficult as I had recently married and started a family, but I had prepared financially to follow my dream. I am extremely proud of this accomplishment as I continued fulfilling my reserve firefighting duties at the XYZ Fire Department where I've been taking in as much knowledge and experience I possibly can (talk about experience here). Working with firefighters and management personnel, I understand the culture of this department and I believe I fit in very well. I give you my word I will give you 100% if given this opportunity."

REMEMBER, YOU MUST SELL YOURSELF!

Situational Questions

Some agencies use situational questions which are asked to test your logic and reasoning. There are no right answers, but there are definitely wrong answers.

Arrest and control technique classes are a great way to become comfortable with the tools you will utilize. Working in the security and/or loss prevention field will increase confidence in arresting miscreants.

Tools Available to You While on the Streets

Police unit and its associated safety features

Firearms

Helicopter

Canine

Do not be afraid to use commands and force when necessary. After all, that is a very real part of the job.

Officer's Situational Oral Questions

Question - You are dispatched to a domestic violence situation and there is a man in the back bedroom holding his wife and child at knifepoint. What would you do?

Answer – The panel is not looking for you to be an expert in police tactics; what they are looking for is to see how you think under pressure and how logical you are. First, you need to understand that these situations are real and they happen all the time, and that these questions are not unreasonable. It is easy to solve this crisis situation if you go step by step. To develop a public safety mindset, begin thinking what you would actually do, which will also assist you in handling chaotic situations.

The first thing you would do is call for back-up, then paramedics to stand by in case of an injury. You would attempt to establish rapport with the suspect and engage him in conversation and possibly call for a hostage negotiator. Keep a safe distance, so as not to place yourself in danger or increase the danger to the hostages. To shoot or not to shoot is a personal question that may change with each variable you are given, but you need to be prepared to shoot the suspect in order to stop the threat. Are there other means of lesser force that would solve the problem, such as a taser or bean bag weapon?

Continue to evolve with the problem and be creative. Ask dispatch to run a check of the address or his/her name if you know it. Is there a history of mental illness or a primary care physician who could be called? Is he on medication that we can get him or trick him into taking?

Remember, the street is not sanitary, and you will on occasion need to use force. Do not be hesitant when you do; use it decisively and effectively and the situation will be resolved.

The Highest Level of Intensity

Jail

It is important for you to understand the ultimate authority of a police officer position. You will be taking liberty away from another and must always act in an ethical manner. This is a modern intake facility at a city jail. From this point the inmate will be transferred to county jail or a state prison depending on the length of the sentence.

Jail Cell

Prison

The following are some oral interview questions from many years ago. In no way, shape, or form are we giving you questions that are currently being used; we are trying to assist you in developing a certain mindset that you must have to succeed in public safety. These multiple scenarios will help you open your mind and determine what level you are at and where you should focus your continued studies.

Question - You stop a vehicle for running a red light, and as you approach the car you realize it is your mother. What would you do?

Answer -This is a tricky question for some applicants. They are torn by what they would actually do and what they believe the oral panel wants to hear. The answer should be what you would actually do. Tell your mother the consequences of running a red light and the people she can hurt by driving recklessly and tell her goodbye. Believe it or not, there are some applicants who say they would give their mother a ticket, which means they are either lying or had a bad childhood. These applicants would have trouble in the psychological examination or are trying to placate the oral panel.

Question - Your partner confides in you that there is an underage runaway female staying with him until she can find another place to live.

Answer -You must advise him to bring this to the attention of a supervisor and tell him you will be contacting your supervisor immediately. It does not matter what he asks of you as you are now in a position to protect him, yourself, the department and the runaway. You must do the right thing regardless of your relationship with your partner.

Question - You are on duty in full uniform and walk into a pool hall bar/restaurant to use the restroom. On your way out, a large male biker steps in front of the doorway, blocking your path and tells you the only way for you to get out is through the back window. He is holding a pool cue and seems confident in his actions.

Answer – Let's take first things first. Initiate a friendly conversation and attempt to gain voluntary compliance, and while doing this, assess whether or not he is armed (including the pool cue), has friends with him and has any physical weaknesses in case of an attack. Use your radio to call for back up, but the panel will tell you it is inoperative; you should adapt and use all available resources such as the business phone, cell phone or even request to use another customer's phone.

You continue to converse with the subject and he still refuses to move. What will you do? Some applicants pull their handgun and then they are stuck. How long do you point a weapon at someone before you finally decide to put it back in your holster? Newsflash: Not everyone complies with your orders, even when you're pointing a gun at them. Will you try and take him on physically? How big are you? How big is he? Does he have any martial arts training? Do you have any martial arts training? If you do, will your training protect you against an attack with multiple assailants while you're grappling with this subject? Begin thinking about these types of situations to truly understand your personal skills and abilities.

This is a difficult situation in which you may be forced to swallow your pride. You may very well crawl out the window and come back with the rest of the shift. Ultimately, you'll return, arrest this subject and win this encounter. We always win, we have to win...

Firefighter Oral Interview Tips
Fire Captain Stephen Horner, a veteran firefighter with the Santa Ana Fire Department, has interviewed over a thousand prospective firefighters. His advice is simple yet powerful: Look sharp and be prepared for a good opening and a good closing statement. Smile and speak to the panel. Do your best to connect with them. Understand the culture of the agency you are testing for. You should have visited their facilities prior to the interview and be able to explain what impressed you and why you would be good for their department. If you did your homework, the board will see everything they want in a candidate right in front of them!

The most common problems are the candidate thinking they already have the job and showing up unprepared. Also, the better prepared you are the less nervous you will be.

Fire Situational Oral Questions
Question - While your partner is performing a medical aid; you witness a mistake of medical procedures. This caused obvious medical discomfort to the patient. What will you do?

Answer - Ask your partner what happened first. There are many things that won't make sense until you hear the explanation or gain enough experience to understand what occurred. Not assuming your partner did something wrong will foster a good working relationship, in which you both can improve your lifesaving skills. If your partner did something wrong, whether or not it created medical problems or liability to the department, you and he should speak with your supervisor to let him/her know what happened. This will give your supervisor the opportunity to mitigate any issues that have come up regarding this situation that you may not be aware of. It may also encourage additional training at the department that would benefit both the department and community.

Question - You are an off-duty firefighter eating at a restaurant with your family. You notice it is overcrowded and the rear exit is chained and locked shut. What are going to say or do if anything?

Answer - You are sworn to protect the public and this situation is definitely a danger to their safety. Contact the manager and advise her of your concern as a private citizen and customer. Perhaps the manager has a perfectly good explanation, such as they are working on the underground electrical lines directly outside of this exit, or they are using the alternate exit through the kitchen and it has been approved by the fire chief. If the manager gets argumentative with you, which is highly unlikely, identify yourself and tell him he needs to correct this fire hazard immediately and then follow up with inspectors the next day. You need to protect the public even if it may mean you will be despised by the manager.

Question - You are a firefighter and are on the scene of a medical aid and you see your captain involved in a physical altercation. The police are en route to the scene code three. You understand if you get involved you may lose your position as a reserve firefighter. What would you say or do, if anything?

Answer - Your captain is obviously in danger and you must do what you can to assist him. You do not have time to ask the captain why he is involved with this subject. Your trust in him and his position tells you he is doing the right thing.
NOTE: You should also be able to answer any questions involving the acronyms for fire and the nomenclature of each particular acronym.

Question – You are in the fire station and you see a six-pack of beer with one missing in the refrigerator. What will you do?

Answer – You may feel this is a trick and opt to report it to a supervisor and your chain of command immediately. This is not a good answer as you haven't researched or inquired about anything. Go step by step and the situation will most likely be resolved with a logical explanation. First, look around. Is the beer being used in cooking, shampoo, etc.? You may hear someone talk about the beer that added flavor to the stew they made last night. Don't jump to conclusions and think about the likelihood of a fellow firefighter risking his career to drink on duty. If he had such a problem, why were there still five beers left in the refrigerator?

If you cannot figure it out, ask someone and you may get such a simple explanation you'll laugh at what you feared; however, if a fellow firefighter is drinking on duty he is placing a huge amount of liability on himself, the department and his coworkers. At this point you would notify a supervisor and allow him to investigate the situation. The panel may then ask you, "what if the supervisor told you to mind your own business and keep your mouth shut?" At this point, you must go above him in your chain of command and explain the situation. You may be treated like a pariah by other firefighters, but your ethical standards will be well known throughout the department.

During the situational questions you will have the opportunity to display your analytical skills and discuss the research you've conducted on each subject. You'll need to demonstrate your thought process as you speak. Don't make the panel guess what you're thinking!

Research

Most likely the next question you will be asked is what you know about the particular agency for which you are applying. You should know the city and state, the demographics, and the political environment, including the chief's name and the department's philosophy. You have many resources at your fingertips (i.e., Internet); use them! If you have an opportunity to visit the department and/or go on a ride along, you can incorporate this into the interview. Name-dropping is extremely effective if it is legitimate and you actually know the person. Warning: Nothing makes a person look sillier than name-dropping with no other reason to bring the name up and attempting to ingratiate themselves to the panel. If you use someone's name, make sure to have a solid reason behind bringing it up.

Closing Statement

You will almost always be given the opportunity to provide a closing statement. This is the time to tie up any loose ends that may have happened during the interview. If the panel is still unsure about passing you, a good and honest closing statement might win them over. You want to recap your accomplishments and leave the board with a good feeling. The statement should be approximately 30 seconds long. Be careful not to ramble, as you will experience what is called diminishing returns. We have all heard people speak for no longer than a minute and were bored to death. Conversely, think of someone who you could listen to forever and never get bored or disinterested. Be that person!

There is a great story about a famous author that describes diminishing returns. He was in church, listening to a preacher give a sermon on why the congregation should reach deep into their pockets for some much-needed repairs to the church. After a few minutes, the author was so inspired he pulled a $20 dollar bill from his wallet and was eagerly awaiting the opportunity to place it in the basket. The preacher continued to talk and he put the $20 back and pulled out a $10 dollar bill. A few minutes later, the preacher still on his sermon, put the $10 back and pulled out a couple of ones. By the time the preacher finished his sermon, the author took fifty cents out of the basket for his trouble! We see this happen all the time; an otherwise intelligent, well qualified applicant talks himself out of a job. There is an old slogan which says it all, "you already sold it, don't buy it back!"

Another example of diminishing returns; imagine craving chocolate all day and finally getting a chance to eat some late in the afternoon. The first few bites will be heaven and then you will start to experience diminishing returns; each successive bite will be a little bit less enjoyable than the previous bite. Eventually, you will come to the point where the chocolate will be a negative experience. We want to reinforce this concept, because the same thing happens in the interview! Stop talking just before you begin to lose points due to diminishing returns.

Sometimes you need to consider keeping it simple. If you feel like your interview has gone extremely well and they ask if you have anything you would like to add, an effective closing statement would consist of thanking each of the members of the oral board, telling them you are excited about the prospect of becoming part of their agency and would enjoy working with them in the future. We have had applicants who were going to be hired prior to their closing statement and then go off on a tangent changing the board's decision. On the other hand, if you feel you could have done better in your oral, bring up an "ace" (a positive incident or attribute that you saved for the occasion such as an award, accomplishment or special recognition) that you have not had the opportunity to bring forth. This can be a lifesaver and give you an edge if you need it. Remember, "the prepared applicant is the one that gets hired", so **be Prepared.**

Questioning the Questioners

Do not ask any questions during the process despite what many college professors or advisors tell you. If you feel you must ask a question for whatever reason, ensure it pertains directly to the process, such as, "What is the next step, should I be successful?" Other questions will backfire. In fact, be careful what questions you ask at anytime. You may believe it is an informal setting only to find out differently later. The old saying, "there is no such thing as a stupid question," is not true and in reality is a stupid slogan! Of course there are stupid questions, and stupid questions are a very good indicator of your level of intelligence. If you're wondering what you need to do and want to ask a question, wait until the end as the speaker may answer them during her lecture, or more likely some hapless applicant will ask the question for you. Let the other candidate be the "fall guy." Don't be that annoying individual who asks all the questions (and you all know who we are talking about). He may believe he impresses everyone; I guarantee you he doesn't!

One out-of-state applicant unbelievably asked if he would be getting moving expenses if he was hired. I told him I'd give him a dollar for a bus ride to his local unemployment office if that's what he meant by moving expenses. Needless to say, he was never hired by us. Though this particular applicant was not in a formal interview, as I explained earlier, everything is an interview. My partner and I went to the local academy to present a badge to a recruit we decided to hire. We had the intention of making it special and prepared an informal presentation. At the conclusion of the badge ceremony this recruit asked if he was going to be reimbursed for already incurred academy expenses. We were blown away by this question and then began a very memorable chewing out session. I still remember the lack of common sense this new recruit had and how he immediately turned this positive situation into a negative one. He no longer works for us.

As stated before, it seems the recent trend in the academic world is to teach young people they are supposed to ask questions during and at the end of the interview. Putting it bluntly, nothing could be more wrong in public safety. We won't tell you how the civilian world feels about this practice, but in public safety, if you want a career, do not ask questions during interviews! First, university professors are not considering the generation gap. Yes, 20-year-olds are going for the job, but 40-year-olds are interviewing them. In addition, the interview will usually be conducted by a police sergeant or a fire captain who are not accustomed to answering questions from a subordinate. The interviewers should get to know the applicant, not the other way around, and the interview should not serve as a research tool, it's too late for that! Also, 90% of questions are comments in disguise or have a specific agenda; it is not worth the risk, so don't do it!

Hopefully, you get the point. Do not ask any questions during or at the conclusion of the interview…EVER!

Post-Interview and Feedback

Drive to a safe location and go eat lunch. Write down everything that was asked and what you said in response. Continue thinking about your interview and how you can improve on any future oral examinations

If you fail the oral interview, some agencies offer feedback. Feedback is an overall evaluation of your interview. If you are courteous and respectful in making your request, the personnel representative may give you information and some golden nuggets on how you could improve. The reason for the agency offering feedback is to assist you in your improvement. If you've gotten this far, chances are you are a good candidate who just needs some polishing. It is not offered for you to gather information and use it against the city to file a lawsuit. If this were to ever happen, feedback would disappear for all applicants. It would also be counterproductive for you, as the agency that does eventually want you will turn and run during the background investigation.

If the organization does not offer feedback, you will need to discover what you answered incorrectly by yourself. Ask officers and firefighters the questions you were asked and listen to their responses. The more you speak with the actual practitioners, the better understanding you will have of public safety. Ask them how they would have answered each question during an oral interview versus an actual situation and see if it's different. Ask them to give you additional scenarios and try to answer them without their assistance. Ask for a critique as most would be willing to help you and offer their infinite wisdom; however, do not take what each public safety official says as gospel. Evaluate what they tell you and develop your own mindset when handling situations.

INTERVIEWS
Practice, Practice, Practice

The most important facet of doing well in the oral interview is to practice, practice, and practice some more. Just as you practice anything else, you need to be willing to give a 100% effort, which means wearing a business suit while practicing answering oral questions. We suggest conducting what are called mock interviews several times prior to the actual interview. These mock interviews do not have to be with public safety officials. In fact, you can do them in your own living room with your wife or husband and/or friends and neighbors. We strongly suggest you make it as real as possible by going through all the motions you would go through in an actual interview; after you get done with your giggling it will be effective. Shake hands with your spouse during the greeting and have a set of questions ready that they will ask you. You may have even written the questions, it does not matter; the point is for you to get comfortable articulating yourself to another person and get used to formulating a strategic way of thinking.

If you want to go all out (and you should), set up a video camera facing you from an interviewer's perspective. This will allow you to see what they see, and hear what they hear! You will observe nuances you are unaware of, such as swaying side to side, turning your body and/or avoiding eye contact; you will hear all the "ums, ya knows and yeahs" you say and other bothersome habits you may have. You will see a major improvement between your first and last videotape and will be more likely to be successful in your interview. Applicants sometimes amaze us with their logic; they will spend four years in college, yet fail to practice for even a few days on their oral interview preparation. An even more effective way to improve than using video, depending on your particular learning style, is using a simple tape recorder. The more you speak into it and listen to yourself the more natural your oral interview will sound to the panel.

Oral Interview

We will emphasize specific necessary points on paper, but nothing captures an oral interview as well as being in the same room.

We have put together a comprehensive demonstration of an entire hiring process from A to Z, including interviews with actual experts who have sat on thousands of interviews. This is as real as you can get before your actual interview. They will tell you secrets and the do and don'ts that will make or break you. The video will show the pauses, eye contact, hand motions and many other nuances that can only be seen from a screen.

This DVD will offer additional tips to improve your odds of success.

If you study this book and DVD and give your best effort you will be on the road to success. Our goal is that you succeed and we will assist you in making that happen.

To order How to Pass the Oral Interview book and DVD, visit our website at www.hiredbypolice.com

Background Investigation

If you pass the oral interview, your next step will be the background investigation. This is the most arduous component of the entire process. Your entire life will be dissected and evaluated to determine if you are suitable for the position, physically, mentally and intellectually. And remember, everything that occurs in the hiring process is confidential and will not be disclosed without your approval.

The background investigators are charged with the task of protecting the reputation of their agencies for the next 25 years. They will assess your overall attitude and your openness to criticism. Your attitude and willingness to learn from your mistakes are the best indicators of success. Background investigators have the best interest of the department foremost in their mind, but they also want to see you succeed. If you are not suited for public safety or their particular department, they will tell you and may even point you in the right direction.

For example, an applicant who had a genius level IQ and was in excellent physical condition confided in me that he believed every person in the street could be reasoned with and force would never need to be used. I wondered what world this applicant had lived in for the last 22 years. I had to know if he had what it takes to use force and adapt his views to the violent world of police work. I spent many hours explaining how criminals and officers co-exist on the street. After a few days of discussion, he amicably withdrew from the process. I was satisfied because I know that now he will not die on duty, which is what would have happened to him. This is a quality individual, who just wasn't meant to be in public safety.

It is a background investigator's job to put the pieces of the puzzle together and determine if the candidate fits into the bigger picture of public safety. Background investigators are usually fair-minded and conscientious individuals. They care about their department and community and will do whatever it takes to protect them. Always be respectful regardless of their findings during the background investigation. Remember, they are fact finders, and whatever they discovered is what you did, not what they did!

It is important for you to assess the personality of the background investigator. This is similar to what you will need to do on the street where you will assess people within seconds for purposes of survival. Learn what is important to the investigator and accommodate what she wants you to do. The background investigator should be your whole world; when she says jump, you ask her how high? If she tells you she is missing some paperwork, which is very common, you call her back immediately and let her know you're on the way. That will impress her and encourage her to be on your side. Show the investigator respect and **never** call her by his first name. If she insists you call her by his first name, do it in a professional manner.

Why is a background investigation so important?
Civilian versus Public Safety

The hiring requirements for civilian positions are less demanding than that of sworn personnel. In general, a manager or owner may take a chance on an unsavory person if his work product is above and beyond the norm. What does he have to lose? A few dollars if it goes bad, a customer or two if he decides not to conduct damage control after the perceived inequity. With public safety however, the department doesn't get a second chance.

111

An infamous fire department arson investigator first attempted to enter public safety via a large sheriff's department. He was disqualified in the psychological component of the background process. A nearby fire department did not conduct a psychological examination during their process and the investigator was ultimately hired. What occurred in his career was devastating to numerous families, firefighters, and the reputation of the department. It could have easily been devastating to the sheriff's department if not for their fortunate break during the hiring process.

Consider what that fire department has gone through in the past decade because of this individual. He was setting fires to structures for years, and coincidentally was almost always the first official on scene. He would then investigate the incident and even speak about the cases while teaching other arson investigators throughout the state. Several people died in these fires including a child and a brother firefighter. Let me take the "brother firefighter" comment back as that would give the investigator a distinction he does not deserve.

I think you get the point, if the fire-starter were a plumber, electrician, car salesman, etc., the fires would have been long forgotten, except by and with all due respect, loved ones. Unfortunately, that department will forever be associated with this tragic situation.

Regarding the scandal at the police department you read about earlier in the book; the officer simply did not belong in law enforcement in the first place and was unfortunately pushed through the background process due to a personnel shortage. The background investigation is critical in determining who can and cannot handle the stress of a job in public safety. If an officer or firefighter is deemed unethical or rogue, the first step administrators take is review the background file.

Other officers began their career trying to do the right thing and found themselves sliding on the proverbial slippery slope caused by their "minor" indiscretions. It may have started with a simple fabrication of

112

probable cause in an effort to put a criminal behind bars; their intentions were to protect the public, but they didn't realize their own failure to abide by the law made themselves criminals. Background experts know the best indicator of future behavior is past behavior.

Higher Standard

Hopefully you understand that by entering public safety you are held to a higher standard than any other profession in the world. A jury in most cases will believe what you tell them just because you are a police officer or firefighter. A famous attorney said something about being a police officer that I will never forget, "A police officer is the most powerful position in our country." Think about it, you can pull over anyone you please and arrest anyone you deem a miscreant. There will be people you despise, and people who you know are committing crimes, but you will continuously check your personal ethics to ensure you have legitimate probable cause to stop someone. When you do make an arrest, you could easily abuse the prisoner and probably get away with it, yet you still treat him fairly. Only you will know if you are telling the truth and staying within the law. As the old saying goes, "with great power, comes great responsibility." You now have some insight and an understanding of why the hiring process is so stringent.

Character

Some people believe you either have character or you don't. That is not necessarily so; you can start building your character today in everything you do. Stop participating in gossip, negativity and sabotage; make an effort to change. Be better tomorrow than you were today. The following pages contain the Code of Ethics for both law enforcement and fire. These hallowed words are the guideline background investigators use to hire police officers and fire personnel. They are looking for loyal, hardworking ethical young men and women who have the character to withstand temptation and compromise. Corruption is easy to fall into and can start with the slightest breach of your oath.

Law Enforcement Code of Ethics

As a Law Enforcement Officer, my fundamental duty is to serve mankind; to safeguard lives and property; to protect the innocent against deception, the weak against oppression or intimidation, and the peaceful against violence or disorder; and to respect the Constitutional rights of all persons to liberty, equality and justice.

I will keep my private life unsullied as an example to all; maintain courageous calm in the face of danger, scorn or ridicule; develop self-restraint; and be constantly mindful of the welfare of others. Honest in thought and deed in both my personal and official life, I will be exemplary in obeying the laws of the land and the regulations of my department. Whatever I see or hear of a confidential nature or that is confided to me in my official capacity will be kept ever secret unless revelation is necessary in the performance of my duty.

I will never act officiously or permit personal feelings, prejudices, animosities or friendships to influence my decisions. With no compromise for crime and with relentless prosecution of criminals, I will enforce the law courteously and appropriately without fear or favor, malice or ill will, never employing unnecessary force or violence and never accepting gratuities.

I recognize the badge of my office as a symbol of public faith, and I accept it as a public trust to be held so long as I am true to the ethics of the police service. I will constantly strive to achieve these objectives and ideals, dedicating myself before God to my chosen profession...law enforcement.

Firefighter Code of Ethics

As a firefighter my fundamental duty is to serve humanity; to safeguard and preserve life and property against the elements of fire and disaster; and maintain a proficiency in the art and science of fire engineering.

I will uphold the standards of my profession, continually search for new and improved methods and share my knowledge and skills with my fellow firefighters.

I will never allow personal feelings, nor danger to self, deter me from my responsibilities as a firefighter.

I will at all times, respect the property and rights of all men and women and the laws of my community.

I recognize the badge of my office as a symbol of public faith, and I accept it as a public trust to be held so long as I am true to the ethics of the fire service. I will constantly strive to achieve the objectives and ideals, dedicating myself to my chosen profession--saving of life, fire prevention and fire suppression.

Unfortunately, officers and firefighters don't always uphold the code of ethics and on occasion embarrass their comrades. It is easy to make mistakes and get embarrassed as you are entering a high profile profession that is constantly being watched; and there are temptations to those without the utmost integrity. A man named Lord Acton is often quoted in relation to public safety, "Power corrupts and absolute power corrupts absolutely." Corruption usually starts out seemingly harmless, but eventually grows out of control. It could be a fire inspector getting pay-offs to approve substandard plans or an officer taking a bribe for looking the other way at certain times of the day. A common belief among hiring officials is that there is only a small percentage of public safety officials who would stoop to this level; they also believe they can make that percentage even smaller with a thorough background which delves into the character of the candidate.

Ethics

The following is a real life example from a large police agency: A high producing narcotics unit was breaking records for confiscated drugs and money seizures on a weekly basis. One Friday afternoon after a particularly difficult week, they used a small percentage of seized money to buy pizza for the office. This was the beginning of the slippery slope of graft and corruption they found themselves sliding into. The next time using the money became a little easier and eventually the entire team was using drug money to purchase new jet skis, boats and other toys they felt they deserved. They were all caught and prosecuted and became the poster boys of disgrace. If there had been only one strong officer in that group with integrity, would it have happened? Now, because no one stepped up to say it was wrong, there are several officers in prison. These officers were probably well-intentioned hard working employees trying to make things better, and then for some reason dropped their guard and yielded to temptation. They became a disgrace to the uniform and in particular, the agency. As a former background investigator, I can guarantee you the investigators that hired them were also affected with deep disappointment on many levels. Background investigators are the guardians at the gate; they determine whether or not you belong in this career field and if you and the department are right for each other.

Background Networking

In the past, police and fire agencies had very little idea of what was occurring at other agencies and had difficulty cross-referencing applicants and their files. That is rapidly changing as computers are being utilized to form a data base of disqualified applicants. There are new companies with all types of information available to any department that pays a monthly fee. There is also uniformity developing as background investigation conferences and seminars are offered on a quarterly basis.

If you lie to obtain a public safety position, there is a good chance it will be discovered, and at the very least even if you survive the process, you will feel unworthy. Your feelings aside, it is not a good idea to attempt to fool an agency for many of the reasons you have read about in previous chapters, but most important is the consequence of perjury. Eventually being fired and potentially prosecuted if you were actually hired under false pretenses.

Background Orientation

There are a couple of different ways agencies handle background orientations. One way is individually, which makes life difficult for a background investigator as they must explain the same paperwork over and over again. More common is a group orientation, which is much easier for everyone as there will be less confusion and all candidates are guaranteed to get the same instructions. Do not be late as this may be the first time you meet your background investigator and you want to make a good first impression. Ensure that you are wearing your best suit and have a pen with you. Do not ask stupid questions. The background investigators have been doing these orientations a long time and have choreographed them close to perfection. Everything you need to know, they will tell you. Take copious notes and pay close attention as they may throw a test or two at you, such as "Question #17 requires a dollar figure next to it." They then wait to see if someone asks a stupid question demonstrating their cognitive ability and/or wait until you turn in your background package and see if you followed instructions.

Be prepared to write an autobiography or a synopsis of your life and/or experience. Make sure to use proper grammar and format. You are being evaluated on the way you write, as police reports or after-action fire reports are important parts of the job. We suggest you have an autobiography that you have practiced writing many times. Make sure the autobiography is well balanced, covering all phases of your life including growing up, education, family, military and your current status and future goals.

Preliminary Questionnaire

You will be given a speech on integrity and the consequences of lying and then asked to fill out a Preliminary Background Questionnaire. The philosophy is sometimes described by the following motto "If you lie… you die!!!" Remember, if there is a problem in your background it may be overcome by explanation or time, but if you lie about it the current and undeniable issue will now be **your integrity.**

This questionnaire asks about domestic violence, tickets, accidents, warrants, drug use, alcohol use, disciplinary action and sexual harassment. Keep in mind, if you lie on this form or at anytime during the background process you will be barred from employment at that particular agency for life and any other agency that discovers you lied. If you followed our advice earlier in the book and spoke with a background investigator if you had any concerns, you will have no problems. Don't be embarrassed about your situation as experienced background investigators have heard everything.

After you complete the form, the investigator will take you to a private office and discuss any areas of concern either of you have. She will most likely let you know at that time if employment at that department is possible. Remember not to make a bigger issue out of something than it is. Many times, worrying about a non-consequential incident is indicative of a deeper issue and will raise red flags.

Personal History Statement (PHS)

You will hear all background investigators refer to this as the "PHS." It is a large packet you will need to fill out prior to your appointment. You can start now by getting addresses and current phone numbers of your immediate family, friends, co-workers, supervisors (current and former). Download the PHS from the POST website and complete the packet ahead of time. Agencies will utilize different formats, but the information will be the same. You can make your background investigator's life much easier if you list home, work and cell phone numbers; if his life is easier, yours will be too. Once again, if you pre-planned and had the foresight to get your documents ahead of time you will set yourself apart from the rest of the group. When the appointments are offered, take the first one as it will impress your background investigator.

Appointment

Wear your best suit and be prepared. Be on time and understand you are still on an interview. Don't get too comfortable, be formal and use titles until told otherwise by each individual, and even then, still use the title. You are demonstrating to the hiring staff that you will do well in the academy and would be a good representative of the organization. Your paperwork should be perfect and they don't want to hear that you couldn't locate something. If you could not find it, document the steps you took and the effort you made in your attempts. Remember they are looking for the "can-do" attitude, which leads to success. If your background investigator calls you and says he needs a form signed or a specific piece of paperwork, obtain it promptly and call him immediately to advise of your progress. The harder you work, the harder he'll work and the faster you'll be hired.

Watch your filler words when communicating with your background investigator. Responses such as "ok, uh-huh and yeah" are fine occasionally, but it gets old, just as it does for the oral panel when said too often. To guard against these filler words, have your family and friends correct you when you say them. You will be surprised how often you use them. When speaking, try to curb the use of "ya know," a very prevalent expression with the younger generation and bothersome to the older ones, who will likely be those making the decision to hire you.

No Surprises

Notify your past employers, supervisors, family and friends that they may be contacted by the department and that you would appreciate their support. It is respectful to them and most people take it as a compliment that you believe they are important enough to use as a reference. While you have them on the line, ask for their best contact numbers such as extensions or cell phone numbers. If there was a negative contact in the past, it might be too late to make it right, but be the first to let your investigator know what happened so the issue can be fairly evaluated. If you do not bring it up it will look like you were concealing the issue, which will reflect poorly on your character.

This brings up the issue of ex-wives, husbands, girlfriends and boyfriends. Let your investigator know if there will be any allegations made by anyone from your past. This occurs more frequently than you would think and it is in your best interest to give your background investigator a heads up. It would have been much easier if you maintained a friendly relationship with all of your former partners, which shows a high level of maturity, but we understand that it is not always possible. If there is problems with an ex, explain that to your investigator, but if you were the one that caused the problems (affairs, gambling, drinking, etc.), admit it and be ready to explain how you have matured and corrected the issues. Another way to improve this situation is to have a typed explanation prepared giving the full story of what has occurred and any insight you can provide. In all explanations provided to your investigator use the, who, what, when, where and why format to ensure complete answers. Understand that they are not looking for perfect applicants, because there is no such thing; what they are looking for is an applicant who has learned from his/her mistakes and reached the highest level of maturity necessary to succeed in a public safety career. Also remember, background investigators are issue-driven, not incident-driven. Take responsibility for your actions.

Resolving Background Issues

At your first appointment with the background investigator you will go over the PHS line-by-line and discuss any issues that may affect your future with the organization. He will also go over the polygraph questionnaire that you have filled out. You will be tied to these answers forever, so if you are unsure about anything bring it up now! Ask questions rather than risking an omission or the appearance of deception; remember, **"if you lie, you die."** It never fails, no matter how much we emphasize telling the truth, there are a continuous number of applicants who lie about something so insignificant that it would not have disqualified them if they had told the truth, but now that they lied, their entire future in public safety is at risk. They will not be allowed to reapply at our department and most likely will not be hired by any local agency. Networking and communication are getting stronger among background investigators. If you are concerned about anything that may come up in the polygraph or psychology examinations, and it has not already been discussed, bring it up now and let the investigator assist you in your preparation. If your investigator makes a suggestion, it would be in your best interest to act on it immediately.

Private and Personal Issues

A surprising number of applicants, both males and females, were sexually molested as young children. They are usually embarrassed and often fail to disclose this information to a background investigator. If you are one of these people, don't be embarrassed, it's not your fault, you didn't do anything wrong. The investigator will handle it in a discreet and professional manner.

If you do not disclose this information, however, you run the risk of showing deception during questions of illegal sexual contact during the polygraph examination. You will also be raising red flags in your psychological examination for failing to acknowledge what occurred. Good things may come from disclosure as the process of healing can begin. Many people believe seeking psychological treatment is the kiss of death. They couldn't be more wrong. Understanding that you have unresolved issues and having the intellect and maturity to deal with them shows you are a true grown up. Most applicants had never told anyone about the molestation and say a weight was lifted off their shoulders and they felt an immediate sense of relief. Do not let this unfortunate situation you were involved in continue to be a burden.

A Sordid Past

Obviously it will be much easier to obtain a public safety position if you did the right things when you were younger. That being said, nothing is impossible if you want it bad enough. Every agency will evaluate your background a little differently. You need to understand the culture of the organization and determine if it will accept the negative part of you to win the new and improved version.

Most educated people understand that the best indicator of future performance is past behavior; and in some instances the only way to mitigate your mistakes are time, distance, perseverance, effort and change. For a great autobiography and a story of how a gang member became a police officer read The Two Badges by Mona Ruiz. She overcame difficulty in her teenage years and is an inspiration to many inner city youths. If Mona can rebuild herself, so can you and it starts right now with this book.

If you have made mistakes in the past do not say, "I'm glad I made those mistakes because they made me the person I am!" This would imply a rationalization that you believe the mistakes you made were a contribution to society. Instead, have the mindset of, "I regret what occurred and have learned a great deal from the incidents. In the past five years I've been able to help others avoid the same mistakes I made by mentoring in the Big Brother At-Risk Youth Program." Hopefully you get the point; acknowledge your mistakes and explain what you learned, emphasizing the improvement you have made. If you made the same mistake more than once, the above statement would be without merit so never repeat a mistake! Also, when you are making a statement, both in writing and verbally about a negative incident; be concise, as long explanations may make the incident seem much larger than it is. Do not cut and paste the same explanation over and over again in your paperwork as it gives the perception it is more serious than it really is.

Reasons Actual Applicants Were Disqualified:
Poor Judgment by Applicants

An applicant was in the Army Reserve and received orders to deploy to Iraq during the War on Terror. He felt a sense of impending doom and that he may not come back alive. The night before he left, he tried cocaine for the first time. Well, he wasn't killed in the Middle East and subsequently applied for law enforcement. He was disqualified, not so much for using drugs, though that is a disqualifier in most cases, but for his lack of maturity and inability to deal with stressful situations.

Do not misunderstand the point, using cocaine was definitely an issue, but a bigger issue was his extremely poor judgment. How will he react when facing danger on a daily basis? Will he decide to try heroin next? Everything you do and every decision you make affects your future. If you know something is wrong, don't do it!

A background investigation on a firefighter applicant revealed that ten years prior he threw a glass jar at a female in a restaurant resulting in her being knocked unconscious. He can't un-ring the bell; he made the decision to throw it and now he has to live with the consequences. First of all, can you imagine doing that, especially to a female? Should we take someone that has the capability of uninhibited behavior as significant as that? It does not matter how much time has passed. Based on this single incident, I would not want this person treating my family during an emergency. Is it possible to rebuild himself? Yes, with certain agencies and friends at that department, he may get an opportunity to prove that he is not the same person who committed that crime.

This book may encourage or discourage you from a career in law enforcement or the fire service; either way it was the right choice. You must be meant for public safety and it must be meant for you.

A female applicant was upset about her husband having an affair. She worked herself into frenzy and while he was sleeping hit him in the face with a high heeled shoe causing a broken nose. He woke up and realizing he was being attacked ran out of the house. She chased him down the street, throwing wine glasses at him as he ran. We do understand a momentary lapse of judgment such as this, especially under the circumstances, and we may or may not have been able to work through it; however, she didn't tell us about it and it was discovered during a background neighborhood check. She omitted information (a lie!) She will never work in public safety because of it!

Unbelievably, a surprising percentage of people have experimented with bestiality (sex with animals). A male applicant worked on a house cleaning crew and whenever he could get away by himself he would violate the dogs in the house in a variety of ways. This person seemed normal and so did all the other "puppy love" candidates.

An HMO receptionist was collecting cash co-payments from patients and was facilitating lunches for her and her co-workers with the money. She did not have permission from the doctor and actually believed she wasn't doing anything wrong, even after we explained the components of theft and embezzlement to her. She said, "Everyone does it."

An office manager collected money from her co-workers to be used for group T-shirts and used it on a gambling expedition. She had intended on buying the T-shirts with her next paycheck, but her car broke down and she had to use the money for repairs. Her co-workers were rightfully upset and her co-worker relationships went down the tubes.

Numerous candidates thought it was okay to bad mouth their coworkers and supervisors during exit interviews because they were told it was confidential. There is <u>nothing</u> in this world that is confidential. Ben Franklin said, "Three people can keep a secret, if two of them are dead." Stay quiet and gracious and make the best of a bad situation. Never bad mouth anyone, because it will inevitably affect you in some way; even if you are not aware that it did.

One candidate, who was a reserve firefighter, wrote a scathing exit letter to his last department about his coworkers and supervisors. There is only one thing worse than verbally trashing your former department and that is to put it in writing!

One applicant was on vacation in South America three years prior to his application and tried cocaine (one time). He said it was not illegal in that country so we should not hold it against him. His argument was that he is a more worthy candidate because he now truly understands what drugs are all about. I told him that if that were the case, all officers would be required to commit crimes since they are the people investigating them. Bottom line, he had weak character and is not a police officer.

A candidate, who was extremely qualified, was working as a security officer several years ago investigating an alarm activation at a store inside of a mall. His partner opened the cash register and took two twenties out and handed one to our applicant. He took it and said he immediately experienced remorse and planned to tell his supervisor what he did at the end of his shift. Unbeknownst to them, they were being monitored by the supervisor on closed circuit video. He was fired before the end of his shift. He should have stopped this situation immediately, and may never work in public safety because he didn't. Keep your ethical guard up at all times.

A candidate, believe it or not, was on probation for "conspiracy to commit murder." He discovered his family member was involved in a murder and refused to cooperate in the investigation. He was an outstanding candidate in every other way. I respect his decision not to testify against his father, but in doing so he made a decision not to be a firefighter. Based on his thought process, I personally would not trust this applicant to do the right thing.

An applicant was a valet and would take change and small bills from the center console of the vehicles he parked. He was arrogant, carried a sense of entitlement and up until the moment he was disqualified did not believe he did anything wrong.

Timeline of a Background Investigation

The background investigation will take anywhere from a few weeks to a year depending on the depth, intensity, and issues involved. Other factors could be the staffing of the personnel section and the number of firefighters and officers they are hiring. The truth is, the best prepared candidates are worked first, so do everything we told you and you will be one of those! You should check in with your investigator once a week or exactly how often he tells you to see if he needs anything or has any additional questions for you. This will show him you are motivated without seeming overanxious.

What if you are disqualified?

If you have done something that merits disqualification, you have two choices - accept it or appeal it. An appealing is a bad idea in most cases. The likelihood of you obtaining a position in public safety after the challenge is zero. If you accept the disqualification, you can attempt to resolve what occurred and improve yourself in your weak areas. Maintaining a positive attitude, possibly repackaging the way the issue was presented and a constant focus on your goal will look good to any prospective future employer.

How to Handle a Disqualification

If you are disqualified, or DQ'd as it is called in the business, you need to reevaluate your background package. Do not continue to turn in the same package to other agencies and hope that someone accepts it. If the background investigator is willing to tell you the problem area, you are way ahead of the game. Unfortunately, most will not and you are going to have to make yourself as objective as possible and evaluate your package as though it were someone else's. Your first consideration should be, "are there any discrepancies in the information I provided", which is the most common problem. I would suggest you reevaluate your education (i.e., grades, GPA, number of class withdrawals or F's because you failed to withdraw), employment (dates of tenure listed, reason for leaving or a job omission), finance accounts (status, explanations of problems negative information – verify with a current credit report), drug usage listed and potential information regarding other agencies you have applied at.

Chief's Oral

A chief's oral is one of the last hurdles you will have. Though it's called a 'chief's oral', it will probably be given by one or more of his representatives. There will be other applicants with appointments before and after yours so understand you are still in competition. Do not share the questions with anyone else you befriended in the lobby while waiting to be called. In this interview they will ask you more personal questions than they asked at your first interview. Re-read the section on oral interviews and present yourself all over again. Do not assume they know your background and qualifications. Remember to be personable and likeable as this is your shot. You made it this far, so don't blow it by failing to sell yourself!

Questions in the chief's oral will be derived from strengths and weaknesses discovered in your background as well as personal situations, goals and aspirations. Approach these questions and others with a positive attitude. Questions can be similar to:

1. We see you have had trouble with your credit. How can you guarantee there won't be bill collectors calling the department?

Answer – Yes, it's true that I have had trouble with credit in the past, but I would like to point out that it occurred after I was caught in a corporate downsize which surprised many of us. I would like you to know in the past two years I have had perfect credit, and even though I could have declared bankruptcy like many of my peers chose to do. I have fulfilled my obligations and have paid every debt in full.

2. You've had an ugly divorce. How do we know she won't make other allegations and embarrass you and our department?

Answer – Yes, it's true that the relationship didn't end amicably and she does have some negative residual feelings, but any contact we have had has been initiated by her. I have no reason to see or even speak to her, and you have my word that I will not ever do anything to embarrass this department.

3. Your GPA is only 2.6. Why should we hire a person that does not give their best effort?

Answer – (Be careful with a question like this or it may backfire. It is obvious the panel member is not impressed with your GPA. If you come across that you were satisfied or as if that was your best effort, they may think you are lazy or stupid. Instead, answer it like this): Yes, it's true. My GPA was only 2.6, but I think it's important to note I was carrying a full-load of units, while working full time at the student center and also putting in 20 hours per week at the fire department for my internship. I hope to continue on with my Master's Degree and improve on my GPA.

In almost all cases, when admitting you could have done better and refocus the direction of the question, you can turn a negative into a positive and **"ace"** the interview.

CREDIT

Can you rebuild your credit?

You may wonder why credit is so important. Credit history is often utilized as an indicator of how responsible you are, which correlates to how responsible you will be in a public safety position.

Many applicants attempt to avoid a poor credit history by not having any credit; this is also a mistake! You must establish some credit to prove your maturity and ability to be responsible. Use a gas and credit card and pay on time every month and your credit will be established quickly.

A substandard credit history is an indicator of someone who lives a chaotic lifestyle which could lead to further problems on the job. To put it bluntly, if you had to trust your life savings with one of two people - one with a perfect credit history seeming to be in control of their finances, the other with a record of credit problems and a history of trouble paying bills - who would you leave your money with? That is exactly what happens; police officers and firefighters have access to property and money when the victims are in crisis mode and vulnerable.

As a police officer and firefighter, you will have access to large amounts of valuables and oftentimes hundreds of thousands of dollars in cash. If you have had trouble paying your bills, this easy money could be considered a quick fix and be more temptation in theory than it would on a person deemed responsible with their money.

The good news is even if you have a poor credit history you have an opportunity to turn a negative into a positive by doing everything within your power to correct your mistakes.

What is a FICO?

A FICO score, as it is commonly referred to, derives its acronym from the Fair Isaacs Corporation. It is a mathematical calculation based on many origins of data along with your payment history. A FICO Score is data collected by three major credit bureaus that are paid by lending institutions. The higher your credit score the less of a risk you are to give credit to, and the lower your interest rate will be on money you borrow. It is a way of rewarding responsible people who pay their debts on time. Credit scores are between 350 and 850, with 850 being the highest score possible. Anything above 750 is considered excellent and anything below 500 will make it difficult to obtain a loan as you are considered high risk.

If you have made several mistakes in paying your bills past due or have received judgments against you, it is time to correct those mistakes. Your goal is to have the score above 700, and the only way to make sure that happens is to clean up any poor credit history. The first thing you should do is contact the lending institution with which you have had a negative prior history.

Insider Assistance

Make an appointment with the manager and speak honestly about your goal of entering public safety. Take responsibility for your poor payment history. It is absolutely true that your substandard credit history will keep you from achieving your dream if you do not get help. Humble yourself and ask for assistance; ask what you can do to mitigate the issue. Will they allow you to pay the full amount and issue a letter erasing your negative credit history with that particular business? If they agree, get the letter first, pay off the loan immediately and stay in continuous contact with the manager. An opportunity like this may or may not come to fruition, but there is a chance; don't waste a second to seize it when it does.

Tried and True Advice

There are many websites out there that promise to fix your credit or increase your score. Most of them are scams and you should not fall for their quick-fix promises. You didn't get into trouble overnight and you will not get out of trouble overnight. One trustworthy author who has helped millions acquire wealth is David Bach. He wrote Automatic Millionaire and offers "How to Fix Your Credit Score and Improve it in 60 Days or Less," free at www.automaticmillionairecredit.com. Included in his offer is advice on correcting errors and a sample letter which you would send via certified mail. Fixing your credit is among your **highest priorities** as the odds of getting hired with a poor credit history are minimal.

You can also send a letter to each of the main three credit reporting bureaus stating you have a dispute regarding a late payment on your credit report. The credit bureau will send a letter to the institution, and if there is no response, that late payment would be erased from your credit report. Also, if it is determined the report was made in error it must be corrected within 30 days by law. This will work in some cases, but if it truly was your fault, take ownership and do the right thing. Ensure that you get a letter from the lending institution stating your agreement with them. Have them put in writing when you pay them the agreed-upon sum and follow through to make sure they rescind their negative reports to the credit bureaus. Now that you realize the significance of a negative credit history and a low FICO Score, change your spending habits, live within your means and you will be on the right track.

Another good place to start repairing your credit is with a visit to the Fair Isaac website. Consider purchasing their deluxe service for $44.85, which includes credit reports from the three major credit reporting companies. Contact each agency, or go to www.annualcreditreport.com for further information. You are also eligible to receive a free credit report once a year. If you would prefer to deal with each credit reporting agency individually, go online or call the following numbers:

Equifax – (800) 685-5000 for a free credit report

Experian – (888) 397-3742 for a free credit report

Trans Union – (800) 888-4213 for a free credit report and 800-916-8800 for disputes

Additional Assistance

If you are unable to solve your own credit problems, a great organization to contact is Consumer Credit Counseling Services (CCCS), a non-profit company that provides confidential financial guidance and free consumer credit counseling services. They will contact your creditors and get them to lower or drop your growing interest charges and your minimum payments at no charge to you. They will consolidate your bills. You will make one payment per month to CCCS and they make the agreed upon payments to the creditors on your behalf until you are debt free.

CCCS has helped tens of thousands of people throughout the country just like you to get back on track to financial stability. In addition, they educate you on how to avoid the same pitfalls that got you into the situation and how to stay debt free.

Remember, having credit problems can be an incident and not a pattern. Background Investigators are issue driven not incident driven. Most of us will experience credit problems sometime during our lives, but the question is always, "why did the problems occur?" If it was because you went out and bought a brand new pick-up truck, a boat and a couple of jet skis and only then realized you could not pay for them, there is a problem. Think back to the job dimensions listed earlier in the book. The person with this type of financial problem has shown he lacks good judgment, problem solving ability, and if he refuses to pay his bills, integrity.

Acceptable credit issues could be deemed as unexpected medical bills not covered by insurance, divorce, or a death in the family. Problems that arise due to lack of judgment or responsibility by overspending and buying luxury items such as boats and cars that you cannot afford can create issues of concern. The important consideration for your investigator at that point is what you did to resolve the problem. The proper path is to write letters to each of your creditors maintaining a copy for your file. Advise the companies of your financial situation and your intentions to pay them back as soon as possible. Most of the companies will be cooperative, but even if they are not you are still demonstrating positive behavior (i.e., ability to confront and solve problems, which are two of the job dimensions used to evaluate applicants). Always attempt to make arrangements that are acceptable to both parties, and remember, they negotiate with people all the time - you are not alone.

Increase Your Odds!
What Can You Do Right Now?

Distancing the Bad Element

Do not place yourself in a precarious situation. An example of this is going to a party at a friend's house whom you know is trouble. If you have any troubled friends and/or friends that commit criminal acts and you haven't already ostracized them, you need to do it now! I don't care if you're 16 or 36, immediately make them part of your past. You can't be in both worlds. They chose a different path. They are not like you, and if you want or need to be "cool" and stay in contact with them, maybe this profession is not for you! If you try to salvage the "friendship," you are asking for major problems. Make a statement to the hiring officials that unsavory people have no place in your life and you will not tolerate illegal behavior.

Family Members on Parole/Probation

If you have family members on parole, probation, or in jail, it would be best to distance yourself from them as they do not have the same ethical and moral standards that you do. The question is, "can I visit them in prison or jail?" The answer is, of course you can; however, you need to understand you will have to answer for your actions. Why are you in the prison visitor system?

"My brother just got out of jail. I live with him. Can I still be a police officer or firefighter?" The answer is, not if you continue living with him. You need to make a decision on which direction you want to take long before he gets out of prison. Move out and minimize your contact. If he turns his life around and proves it over a significant period of time, only then should you reevaluate the situation with the help of your police sergeant or fire captain.

Maintaining **any** type of contact will stain your reputation as well as your department's image. Make your commitment perfectly clear to everyone - family, friends and mostly yourself.

Negative People

Another bad element you may not perceive as bad are negative people. Negative people affect the way you live your life, and if you are around them too much you will start to think and act like them. Your mindset becomes "everything seems to go wrong" and it eventually becomes a self-fulfilling prophecy. Positive people will begin to avoid you personally as well as professionally. When you come to mind they will think of you in a negative light, which can translate into a non-recommendation during the background investigation.

Character

Character is the most important trait a public safety applicant can have. So give up the parties and drama, the friends with attitudes, the friends who create problems for you, and above all maintain your dignity and character. Living a clean life when you are young might have resulted in you being teased about the way you decided to live your life. Some would call it boring and uneventful, and some may have even tried to embarrass and sabotage you because of jealousy. Trust in yourself and don't worry about people like that; high moral character pays dividends in the background investigation. Your career in public safety will provide experiences some could only dream of. Society will hold you to a higher standard and may even place you on a pedestal. But remember, along with this trust comes great responsibility and expectations.

Renewal with Friends and Coworkers

Start to develop a good rapport with coworkers and your supervisors. If you have had a rocky past with them, do whatever you need to do to repair these relationships. Perhaps you have not been as loyal to your boss as you should have been. Apologize and subordinate yourself to him/her. We will be the first to tell you that some bosses and peers just cannot be pleased, and all background investigators know that; however, most can, so take it upon yourself to establish a solid and positive relationship with you all fellow employees.

The most important thing you can do for yourself is to act professional at all times. A background investigator will be contacting your boss and co-workers and ask their opinion of you. They will be asked, "Do you trust her to patrol your neighborhood when you are sleeping?" or, "Would you trust him to administer medical aid to your sick child?" These hard-hitting questions usually elicit a powerful response, either for you or against you. Work with your supervisor and peers as if you are trying to answer those questions.

Don't Forget Your Allies

Do not assume your best friend will say great things about you to your background investigator. On the contrary, he or she knows you better than anyone and your investigator will spend the extra time necessary to gain truthful information. Remember, your background investigator has spent his career interviewing parolees, child molesters, rapists, murderers and many other dregs of society. He will obtain truthful information despite what the interviewed people tell you! Some background investigators are surprised at what best friends divulge when interviewed. You need to work just as hard to convince your friends as you do your coworkers that you can be trusted in public safety. You will probably never know what was said about you, but if you get hired it could not have been too bad.

136

Pay attention to the following tips as they are vital to success in the public safety hiring process: Never quit a job without giving two weeks notice, and never leave a job with a negative attitude. Sour grapes will be your downfall. Do not tell your boss, supervisor or any co-worker what you actually think of them if what you think of them is negative. When you are leaving a position, obtain a letter of recommendation. Years later when you are applying for a position, your supervisor may no longer work there or the company may be out of business. A letter of recommendation on company letterhead is a lot more impressive than the excuse, "I can't get anything because the company is no longer in business."

The Bad Boy Look

Many of you are trying to be "hard" and you know exactly what we mean: The baggy pants, bandanas, tattoos, gangsta rap and a hard edge voicemail that has you talking like a rapper or gangster, sounding ridiculous. Delete any type of sexual, gang or offensive innuendo on your <u>MySpace, Facebook or Twitter accounts</u>. Grow up and prove to yourself and everyone else that you do not need their approval. Besides, and let us make this crystal clear, you are not hard, your efforts to look like a thug are laughable and you look and sound idiotic. When you become an officer or firefighter and have actually been involved in near death situations and have seen things human beings should not see...then you'll be hard.

Stay within the established decorum of your job title and dress for success; do not be one of those who wear the long shorts and knee-high socks. That particular look actually started in prison and young people today consider it cool. The people that hire you do not consider it "cool" and will send you packing out of the hiring process.

Also, if you're a male and wear an earring, nose ring or any other protruding device, you need to understand how unprofessional you look to most of society, especially to the ones evaluating you. A male applicant who wears earrings is unacceptable to most hiring officials. You will never know if it affected your interview, but a hole in your ear or other visible piercings are guaranteed to negatively affect your score. It will also keep you looking for a job. We have had numerous young people tell us "it isn't fair." Frankly, we agree with them, but it is what it is, and the conservative persona of public safety employees will not change.

Also, do not get tattoos that are visible as many law enforcement and fire agencies have recently implemented policies against visible tattoos. Current employees with tattoos are grandfathered, but prospective employees will be eliminated. The tattoo ban has been tested by litigation in federal and state courts and has been upheld. Why take the chance when perception is so important in how the public reacts to us as peace officers and firefighters? Having visible tattoos could make the prospective hiring agency look harder for another issue in your background.

What if you foolishly injected ink into your body prior to reading this book? There are tattoo removal clinics everywhere and it should be no surprise their business is booming; however, understand you alone tainted your body. You may now have to go through the removal process which is painful and leaves a scar. You must decide if that particular agency is worth the tattoo removal consequences. If you choose to keep the ink and jewelry, there are agencies that will overlook it, but they are becoming fewer by the month, so you better hurry!

Smoking

There are absolutely no redeeming qualities to smoking cigarettes. It makes you, your clothing and your breath smell bad. Non-smokers will smell you when you enter the room and it will not earn you any points during the oral interview. Also, many agencies have implemented and require you to sign a non-tobacco use policy agreement upon employment. Most police officers and firefighters have a disdain for smokers; it may not be fair, but it is accurate. Let's discuss health issues. Tobacco causes cancer and will ultimately kill you and secondhand smoke will kill your loved ones. Do yourself and everyone else a favor and stop immediately. If you cannot be convinced to stop smoking in order to successfully complete the hiring process, at the very least do not smoke the day of your interview.

Jobs, Past and Present

Ask yourself these questions: Are you a good employee? Are you loyal to your boss? Do you gossip? Are you a busybody? With everything you know about yourself; would you hire yourself? Do you really care about the organization's success, or is it just a paycheck? Think about these questions and whether or not you have shortcomings in any of these areas. If you do (and we all do), fix them right now!

Let's develop a plan together to improve your performance. If you are a substandard employee, make a decision to change that immediately. From this point on you will be the "go-to guy or gal." You will be the employee who every supervisor can depend on. This is not difficult to do, but you will have to make an above-and-beyond effort. First, be happy that you are employed and earning a living. There are millions of people throughout the world that would love to be in your position. Next, do not complain when you are assigned a task. Take it on with relentless zeal and ask for more responsibility when that task is complete.

Stay away from negative people, and if they hang around tell them about your new approach and that they will need to either join you or stay away from you. Show up to work early and leave a couple minutes late. What's five minutes anyway when it can improve your work ethic and reputation? Everything you do must be quality; do not settle for less and your performance will be noticed. There is one problem with setting high standards and that is you will have to maintain them, so do not take this approach unless you plan on staying the course!

Improving a Bad Situation

What if you made a mistake at your job and do not know how to recover from the incident? If you are still with the same company, rehabilitation is a possibility as long as you have a plan.

On occasion, we all make stupid comments or confide in the wrong people and we suffer dire consequences. Sometimes what we say 'gets legs' and we may not be able to recover from them. For example, an applicant who was going through the background process was working at a large bank.

While I was interviewing his coworkers I spoke to a black female. She stated the applicant acted inappropriately a year ago when he asked her, "Why do Blacks like chicken?" He followed this question with a laugh and walked away. Try to imagine her disgust; not only was this inappropriate for any setting, it displayed a definite lack of maturity on his part. She became visibly upset, and then he got nervous, but still never apologized. Obviously he was disqualified as he was unfit to serve the public based on several racial and gender bias factors. It is unknown if he was a racist, a joker, culturally insensitive or just stupid and/or all of the above, but what I do know is with this one comment he will have difficulty ever working in public safety. If he had a chance to rebuild himself it would have started with an immediate apology and followed up with discussion and understanding.

Any type of racial or sexual joke is inappropriate and not worthy of a person holding the public trust. If you have done something inappropriate that you have left alone and ignored for fear of stirring the pot again; to 'let sleeping dogs lie' and to not create a second round of drama when there was none. It may be judicious to do so, and there are times that the common-sense approach will let the incident disappear forever; but take note of this, if you remember the incident, so do they! It is very possible they will be interviewed and that situation will come up. Ask yourself, if they are interviewed would you be concerned? If yes, do what you can to minimize the damage and take every opportunity to mitigate the consequences and make it as painless as possible.

Internships, Fire Cadet, Police Cadet, Reserve Firefighter and Reserve Officer

These programs are an exceptional method of entering public safety and obtaining experience if you are not quite ready to take on the responsibility of being a sworn public safety officer. This opportunity can either help or hinder your future. First and foremost, know your place. Remember the food chain chart taught to you in the 8th Grade and understand you are on the bottom of it. You need to be subservient and humble, quiet and respectful, studious and helpful. If you follow those rules you will make friends and possible allies when you apply for a public safety position. Wind hose, take care of paperwork and answer the phone at the fire station. Do the little things that veteran firefighters would rather not do.

Conversely, if you are cocky and/or arrogant you will get the opposite effect. You will have sabotaged yourself and your dream will be over. The most important thing to remember is that you are not "one of the guys" yet, no matter how much they include you in events or discussions. Don't fall into a comfort zone and act as if you are part of their exclusive club. You haven't earned their respect yet. Ironically, that is the very attitude that will assist you in becoming one of them!

Martial Arts

One thing you should do immediately if you are serious about law enforcement or fire safety is to join a martial arts studio. It will prepare you for public safety both mentally and physically and for violence on the street. As an officer you will be faced with confrontations on a daily basis, and as a firefighter you will also need skills to restrain violent people. What type of art you choose to study is up to you. There are many different theories of what works best for public safety officers, and that will vary depending on with whom you speak. Ground fighting (grappling) has gained enormous popularity the last few years due to the television exposure of the "Ultimate Fighting Championship." This is important, but not the initial type of training you should do.

For expert advice on this subject, I spoke with Sergeant Joe Kahapea who has over 35 years of police experience. Joe is well-known in the martial arts community as well as a leader in developing "arrest and control techniques" throughout law enforcement. He has instructed recruits and sworn officers for nearly three decades and has studied the "Lua" style since he was a small boy growing up in Hawaii. His advice is to look for a practical stand-up street fighting art, which also teaches ground techniques.

Joe recommends Aikido for beginners who have a desire to enter public safety. Aikido places emphasis on hand and arm locks, which correlate closely to law enforcement arrest and control methods; however, there are many styles to choose from, such as Shotokan Karate, Kung Fu San Soo, Kempo, Krav Maga and many others. You should choose a style that fits your needs, ability and personality. Talk to the master instructor and tell him what you are interested in learning and he will let you know if it is a match. Also, many colleges offer arrest control techniques as part of their curriculum.

Now keep in mind just because you train in martial arts it does not mean you are guaranteed victories on the street. There are so many factors that determine whether you or the suspect wins the confrontation. The main factor being how much the criminal wants to defeat you along with his rage and purpose. Rage is difficult to deal with whether you have training or not. Most of the time, if you have trained hard, you stand a good chance of winning the encounter.

Training instills confidence, which is difficult to fake around real criminals. Criminals are like dogs in how they sense fear and become more violent when they believe you are weak. Have you ever walked near a growling dog? If you display confidence and act as if you are not intimidated, the dog will not attack, but turn and run and see what happens. Criminals are the same way when they feel they have the upper hand.

Martial arts will also give you a strategic mindset and a confidence you cannot learn any other way. You will begin thinking about distance, and using your feet rather than your upper torso and hands. Chances are you know someone who is training in some type of martial art, so give them a call and get started.

I began training twenty years into my law enforcement career and regret not discovering martial arts when I was younger. I have personally utilized techniques learned in the studio numerous times in the street. I even utilized martial arts in a situation in which I didn't have to use force and was subsequently awarded a Police Service Medal of Valor. I am convinced that training in martial arts and being confident in your ability actually minimizes the need for force. Keep in mind, if you do need to fight, you must win the encounter every time. If you lose, you will die. Ensure the art you choose is focused on reality-based street combat, which translates directly to law enforcement duties.

Firefighters Are Also Attacked!

If you think you're safe from violence because you are a firefighter, you are very wrong! Firefighters wear a uniform and badge and are a symbol of public trust. I know it seems unbelievable and difficult to understand why someone would attack a firefighter, but the criminal mind is a deviant mind and does not think rationally. They see you as a symbol of authority and will get pleasure from the attack. It makes no difference that you are only trying to help; you are a representative of authority.
Police officers understand why some people hate them as they take liberty away and if necessary use force while doing so. Police and fire are usually spoken about in the same sentence and the danger is never far from each other.

Fire personnel have been targeted on many occasions. One of our own firefighters was hit with sniper fire during a structure fire response at the Los Angeles riots in 1992. Thank God, he wasn't killed. In another incident, firefighters and police officers had just finished a call on a graveyard shift and were standing in a parking lot. A gunman targeted the group of officers and firefighters resulting in a gunfight; one firefighter was struck in the exchange. In yet another incident, a local fire department was hosting a toy drive and was caught in the middle of a gunfight between biker gangs; a firefighter was struck with a bullet and fortunately has since recovered. All prospective firefighters should join your brother and sister officers and start your training!

Continuing Education

To make yourself more marketable, you should enroll in criminal justice classes and seminars. Most of these classes are taught by law enforcement or firefighting professionals. An easy way to find out how to enroll is to call the local police department. They frequently have open spacc originally reserved for sworn personnel who could not make their required training. These classes will indoctrinate you into the public safety system as you will be exposed to what we call "war stories" by public safety personnel who are on the front lines. Real life conditions and scenarios will be discussed and you will begin to understand how the public safety official thinks. These are actual police officers and firefighters conducting advanced training. A number of these classes will be restricted to sworn personnel only as confidential information will be discussed; however, this is rare and most of the classes will be available for you to enroll and participate in.

If you stay quiet and humble you may be allowed into the private world of sworn personnel. They will become more comfortable with you as the day goes on and give you valuable advice. Do not speak unless spoken to and definitely do not share a story as it will be detrimental to your reputation. You will get what is called a "jacket" which labels you with a "who does this guy think he is" attitude. If they do open up and discuss the job in your presence you will have never laughed so hard in your life. Professionalism is only surpassed by humor in their ranks. The funniest moments in my life have been in roll calls and firehouses, which unfortunately would never translate to the civilian world because too much groundwork would be required to explain the humor.

Get to know these firefighters and officers. If you become friends, ask them for a reference and advice on getting hired by their agency. The "good ole boy" network is not as prevalent as it used to be, but let's get real. All things being equal, are they more likely to hire you if they like you or don't like you?

Do Your Homework

Get to know the criminal justice and law enforcement staff at local and career-focused colleges and they will assist you in preparing yourself for public safety. Consider taking a foreign language specifically focused on law enforcement.

These classes are often not put into community service class schedules and it is best to personally contact the criminal justice secretary to see what is available. There will be a fee for the class but it will be well worth it. Remember when you call, be polite but persistent as they will not show you the deference they will show sworn personnel that are trying to register. Remember when calling, that you are speaking with a person you should treat with the utmost respect, and if you don't, you will see firsthand the power a secretary wields as your name may be placed on an unofficial "waiting" list.

Writing Police and Fire Action Reports

You will need a base knowledge of grammar and composition to be an effective report writer and win convictions on your cases. Hopefully you paid attention in your high school composition class; remember when you thought you wouldn't use any of that knowledge and didn't work as hard as you could have? All of those times will come back to haunt you if you do not immediately begin studying grammar.

Everyone that reads your report will make judgments regarding your intelligence, work ethic, investigative ability, writing ability and personality. These people include coworkers, supervisors, managers, district attorneys, private attorneys, public defenders, suspects, arrestees and judges. You do not have the luxury of time like they do; you will not have people review your writing prior to it being turned in, since the clock for filing charges begins ticking at the time of arrest. You will be writing reports in between calls and depending on the amount of violence in the street that night you may be writing it after an exhausting 12-hour graveyard shift. You may have been in court all day and had only a couple hours of uneasy sleep, but it won't matter, because once it's on paper; the report becomes who you are.

The more you learn now about writing police and fire reports puts you a couple steps ahead of the curve when you get to the streets. I recommend you read, read and then read some more. Avid readers usually have a significantly better vocabulary than non-readers and also write much better. Start to pay special attention to sentence structure and specific grammatical markings and your skills will improve tremendously.

If you know a police officer or firefighter, ask them to obtain an incident report which will teach you how to properly format a police and/or arson report. To excel will require practice and effort just like everything else; you need to practice to be good. You will need to write, write and write some more; write about everything you do and format the script as you would a police and/or fire report. Speak with an English teacher and tell her what your goal is. Ask her for extra help and to critique your writing in a law enforcement and fire style. Don't make it more difficult than it is, cover the, who, what, why, where and how of the incident and you will be ready.

Firearms

Contrary to popular belief, most officers are not gun fanatics; in fact, most are the opposite. We are extremely competent with our firearm and practice regularly, but are uninterested in having a conversation about weapons. Citizens will approach uniformed officers and start talking about the nomenclature of an M-16 rifle because they think the officer is a gun nut. They always have a surprised look on their face when they are told, "The only thing I know about weapons is how to shoot them."

You may have a fear of guns or be inexperienced in handling them. Actually, most range masters prefer a recruit who has never held a gun before; this way they can start fresh and not spend time correcting bad habits. If you want to become familiar with firearms, look up firing ranges in the Yellow Pages or online. Chances are there is a range master who also works at a police department. Ask them if they give private instruction at a reasonable price. Do not take instruction from officers or well intentioned friends as they do not always shoot properly. Some colleges may also facilitate this type of instruction.

Phonetic Alphabet

It is difficult to decipher letters on a scratchy radio. The military developed a system to identify letters with words and phrases. This system is universal, and though some departments deviate from exact verbiage, you cannot go wrong learning this as soon as possible. This alphabet is utilized as a cultural language that law enforcement use on a continual basis in conjunction with radio and penal codes.

Public safety uses phrases to describe people and things in the presence of people who do not speak that language. Practice by spelling things and words phonetically and you will soon be competent and begin to understand our language. Listed below is the phonetic alphabet.

Alpha, Bravo, Charlie, Delta, Echo, Foxtrot, Golf, Hotel, India, Juliet, Kilo, Lima, Mike, November, Oscar, Papa, Quebec, Romeo, Sierra, Tango, Uniform, Victor, Whiskey, X-ray, Yankee and Zulu.

Radio Codes

Radio codes are vital to building a base knowledge in public safety. Remember, the more you know before you enter the academy or an actual public safety position, the easier it will be to adapt to the culture of the department.

Visit a local police or fire academy and see if they will give you a radio code book to study, or ask a background investigator or an administrator from the training division for one. Most often, they will be eager to assist a hardworking prospective public safety candidate. Some of the codes are not in use. Ask for the order of importance of the codes to study.

Below are several of the commonly utilized radio codes at most law enforcement agencies; ask any representative of the department you are interested in, if it is what they use.

10-4	Acknowledgement	10-20	Location
10-6	Busy	10-21	Call
10-8	In-service	10-22	Cancel
10-9	Repeat	10-23	Stand-by
10-15	Prisoner	10-35	Confidential
10-42	Pick-up officer	10-98	Completed assignment

If you are focusing on one specific department, you should incorporate map reading and city geography in your research. Don't depend on a Tom-Tom or Garmin; chances are your trainer won't allow you to use them. If you learn map reading and city geography well, it will be one less hurdle you will have to deal with during your training period.

A Big Mistake

Be careful what you say at all times. No one should know you are in public safety. Don't be a big mouth. Stay humble and most importantly stay quiet. Advertising your role as a public safety officer is a big mistake for amazingly simple reasons, such as a criminal preparing your meal or the child molester sizing up your little girl to get even with the last cop that put him in prison.

We see officers and firefighters on a daily basis wearing T-shirts with their logos and department names and slogans on it. News flash, to all the new and prospective public servants - not everyone likes police and fire. Police for the reasons I stated before, and fire due to jealousy. Perhaps someone had wanted to be a firefighter and couldn't, or maybe a firefighter stole their girlfriend/boyfriend, or it could be they think you make too much money. Who knows what actually goes through some of these people's minds.

I know good people who blame firefighters for not saving their relative who had a heart attack. These people forget about his being 100 pounds overweight with high blood pressure and working in a high stress position; yet it somehow is your fault - - you didn't respond fast enough or work hard enough to save him. There is nothing you can say or do to change their opinion. Now, it's important for you to understand if good people think like that, can you imagine what bad people are thinking?

I do know if you remain anonymous you will be much safer. If you are at a restaurant that is robbed and they spot your official looking T-shirt, your danger level has just increased exponentially. If there is a deranged lunatic and he is hurting people, the public will look to you to intervene. Cooks may give you some "extra sauce" to get back at the system, so do yourself a favor, be proud, but be anonymous.

There are a few officers and firefighters chosen to be instructors at police and fire academies. They are charged with teaching and creating public safety graduates. They administer the basic knowledge and establish the principles that will keep you alive in the field. One way they train a new recruit is to make them understand the ramifications of making your occupation known to the general public. Academy instructors are everywhere... in stores, gas stations, restaurants and anywhere else you can imagine.

A recruit in my academy class was at a market telling his friends how easy the academy would be to him. He made statements like, "It'll be a piece of cake; it can't be any harder than Marine Corps Boot Camp." What he did not know was an academy instructor was standing directly behind him. The instructor memorized his face and vowed to teach him humility. On the first day of the academy, he reminded the recruit of his comments in the store and guaranteed him he would not make it through the academy. This particular recruit was probably the strongest physical specimen I have ever seen and the attempts of the staff to beat him down physically failed.

He was verbally, physically and mentally abused almost every minute of the day; he was given disciplinary reports for created reasons and then told to re-write them as he still compiled new disciplinary reports. By the fourth week of the academy he was sleeping about an hour a night and looked like Alfred E. Newman from Mad Magazine. He was a beaten man; the tactical officer won. In the movies, after the recruit is broken down, the tactical officer would rebuild the recruit and form a life long bond. This did not happen and eventually the recruit was worn down to the point of exhaustion; he went some nights without sleeping and when he had nothing left, he finally gave up. I personally believe the instructor should have pulled back at some point; however, the recruit should not have been bragging about how easy the academy was going to be, especially in public. If he stayed quiet and humble he would have easily been the top recruit in the class. I still wonder what happened to him and believe he would have made an excellent police officer.

Academies and staff do not play by the rules, and they shouldn't as society does not play by the rules. If you cannot accept this fact you should find another career. The staff will do everything they can to ensure that you belong in public safety; and if you don't, they will eliminate you. You should rent an older movie titled, "An Officer and a Gentleman" with Richard Gere and Louis Gossett, Jr. In a great scene that mimics the police and fire academy, a drill instructor tells recruits, "I will use every means necessary, fair and unfair, to trip you up."

I have heard of some recruits in general conversation discussing their transgressions when they were younger. I have never understood people that do that; it's as if they think their past gives them some type of street credibility. In fact the opposite is true, it makes them look foolish and immature and their comments may ultimately result in their removal from the academy.

Your comments may even be used to impeach and terminate/dismiss you during and even after the academy is over. During a background investigation I was conducting I learned a prospective firefighter (who was about to be hired) was boasting about smoking marijuana while a teenager; on her PHS she stated she did not use marijuana. She finished the fire academy, but needless to say doesn't work for us and her reserve status was also taken away.

Her behavior directly related to case law, "Snowden vs. Burden and the City of Costa Mesa." A police recruit was hired prior to a completion of the background. The City decided not to send their background investigator to Hawaii where the recruit was previous employed. They later received disqualifying information and terminated her employment. The recruit filed a lawsuit and lost due to her misleading and false statements during the background investigation.

153

Keep in mind that if you lie, not only do you compromise your own integrity, but you place the department in a very difficult situation. If we hire a liar, which shouldn't happen but let's assume it does, the department's knowledge of untruthfulness must be presented to the prosecution and defense when a case goes to trial. This case law is known as "Brady vs. Maryland," ultimately rendering testimony worthless.

Academy Preparation

The best way to prepare for the academy or anything else in life is to find someone who was successful in that endeavor and copy them. Find the police officers and firefighters at your college of criminal justice, fire technology and/or general education classes and initiate conversations. The more you speak with them the better prepared you will be.

Have a Spare

When you go to an academy, you will be issued tools of your trade; police will have keys, handcuffs, baton, a weapon, ammunition and pepper spray; fire will have turn-outs, axes, and crow bars. Both police and fire recruits will have boots, and many of the exercises and practical applications will scuff up your boots. If you can afford it, keep a spare pair of shined boots in your car to change into for formal settings and inspections. You will be saving yourself a lot of grief. Bring extra items of everything you can think of, as your less organized classmates will inevitably forget things.

Get to know the employees at your department's distribution center as you will misplace or lose items. If you can't immediately replace a lost item prior to the academy staff discovering it, you will be disciplined. If your new buddies will work with you, try to have a replacement for every item in the trunk of your car.

154

Wimps Shouldn't Be in Public Safety

The new generation of recruits actually has the gall to complain of injuries and fill out Worker's Compensation forms for minor bumps and bruises during the academy. This new way of thinking in my viewpoint is a disgrace to yourself and disloyal to the organization that has given you your big opportunity. The sense of entitlement that seems pervasive by this new way of thinking has created millions of mentally weak Americans. Refuse to fall in line with that mindset; be mentally tough and it will pay off in the streets.

Recently we hired a recruit who looked physically strong, but I sensed something weak about him and predicted he would not make it in law enforcement. My feelings were confirmed when only three weeks into the academy; the recruit said he was injured. I do not believe he was. What I do believe is he was a weak-minded individual who does not belong in law enforcement.

Police officers and firefighters make judgments within seconds of seeing you. Their lives often depend on their ability to immediately evaluate a situation or person. With this in mind, you need to always be on guard and be on your best behavior, displaying your command presence as well as respect for others. If you have a weakness, mentally, physically or emotionally strengthen that particular area prior to the academy.

Everyone is injured or hurting in some way during the academy. That is part of the test, but you don't quit. In fact, being slightly injured can be a blessing in disguise, as it will show your evaluators you have heart. Do not even think about faking an injury, because you will inevitably limp on the wrong leg or favor the wrong shoulder. You laugh, but this happens! Don't let this person be you and live with the regret and embarrassment. Work out to become physically and mentally strong and be a proud public safety officer!

Meeting Firefighters and Police Officers

Most contact with firefighters and police officers will be positive; however, there will always be a few people in your life, who are bitter about everything and will treat you poorly no matter what you do. Ensure you treat them with all due respect, especially if they are senior officers or firefighters. They have great influence over the way others perceive you. Do not bend over backward, just do what you are expected to do. Remember, you have not earned your position yet, so if a senior firefighter is winding hose, volunteer to do it. You should be doing this anyway for experience and out of respect for the senior crew. Volunteer and establish a good reputation among senior staff.

Reputation

The most respected agency in the nation, like all other agencies, is just one videotaped incident away from a public relations nightmare and civil unrest. Once you are a representative the same holds for you; act as if everything you do is being taped. That is exactly why backgrounds are so important and vital to the good working order and discipline of a department.

Gateway Jobs

Are you a security officer or an ambulance driver? These are common jobs that future police officers and firefighters take when starting out. Do not act like an officer or firefighter, because you aren't one....YET!

You probably have significant contact with these professionals and you should take advantage of it. Be friendly and professional. Take their advice and do not tell them how to do their jobs. Security officers can be overzealous regarding enforcement issues on occasion; sometimes they make illegal arrests and when the officer attempts to explain proper procedures it turns into an argument. The officer who was initially trying to help grows weary of this "know it all" attitude. The encounter ends with the officer demanding the security officers guard card and weapon permit, which often leads to a criminal complaint for non-compliance of some type and he is now unemployed. Do your best not to be this type of security guard. Be eager to impress and learn from the officer; get him on your side and perhaps even obtain a letter of reference.

Security or Loss Prevention is a noble career with many opportunities. Internal theft, known as embezzlement, is a nationwide problem. A good loss prevention agent can save their organization millions of dollars through arrests and vigilance, creating deterrence. Promotions and a six figure income are not uncommon with tenure. Casino security is a growing industry ripe for future private law enforcement leaders.

Ambulance crews perform transports for firefighters on a regular basis and can obtain the same results as security officers, positive or negative. Be positive and willing to help; act as if you are auditioning for a fire position, because that's exactly what you're doing. Be mindful of any negative comments by your coworkers and immediately begin separating yourself from the weaker and malcontent employees.

Do not neglect the affiliate opportunities fire provides. Firefighter equipment manufacturers and sprinkler companies are making millions of dollars, and it seems every other month we see a new life-saving product on the market.

Letters of Reference

Start now with your requests for letters of reference from supervisors, management and anyone who can attest to your outstanding work ethic and high integrity. This will benefit you in several ways. You will need to speak with them to obtain this letter, which will make you more recognizable and give you and the manager a common interest. He may take you under his wing because at one point in his life wanted to become a police officer or firefighter. Your work will now be noticed, and hopefully you do quality work.

From a background point of view, a person is less likely to say negative things about you if he gave you a letter of recommendation. If he did make negative comments about you, he is either a liar, hypocrite or should admit he made a mistake when he gave it to you; people rarely admit mistakes. Now march right into your boss's office and ask for your letter today.

If you have access to high-ranking officials from police and/or fire agencies, it is not time to be shy; it's time to utilize every resource you have to get your foot in the door. If you did not use every potential reference and were ultimately disqualified, imagine how disappointed you will be.

Obtain letters of reference from past employers now. As time passes, they will be increasingly difficult to obtain as people frequently move on to other organizations and are sometimes impossible to locate. When trying to gather documents such as employee evaluations and other paperwork from your file you may encounter resistance. They may state that a court order is required to provide employees with copies of anything. You may have to help them understand the importance of assisting you.

According to state employment law, a former employee has the absolute right to review and make copies of anything in his personnel file. With litigation the way it is now, most people and organizations are afraid to do the right thing. Advise them the background investigator was very clear that these documents were essential in being hired. For their comfort, suggest they contact their legal department to provide a waiver for a written request which you will be happy to sign.

Go into the process of gathering documentation with a positive attitude and do not take "no" for an answer. Start now because many of these processes take time, and having all of this ready will impress your background investigator.

Police and Fire Department Waivers

A typical waiver will be similar to this and will be mandatory to sign:
Waivers can differ slightly in their wording depending on the agency and its legal advisor. The following examples are accurate and actually in use at several agencies. These examples are meant to give you an idea of what you will be signing prior to entering the background portion of the hiring process.

WAIVER

The information you provide in this questionnaire will be used as part of your background investigation to assist our department, and the community which it serves, in determining your suitability for public safety. Completion of this questionnaire is mandatory if you wish to continue in the selection process.

Our communities have the right to expect and in fact demand truthfulness from the men and women who serve as public safety employees. Honesty is expected and required from the very onset of the hiring process.

Your background investigator will not distinguish between lies, big or small. Any perceived or deliberate inaccuracies, incomplete statements, untruthfulness, and/or omissions will be grounds for disqualification.

In the event your background investigation uncovers information that you have or are suspected of having been involved in illegal activities, not only will this information likely bar you from employment here or elsewhere, we will seek charges if the offense is within the Statute of Limitations.

Remember, your responses may be validated by an in-depth background investigation and polygraph examination.

I have read and understand this form

Release and Waiver Police and Firefighter Applicant

To whom it may concern:
I am an applicant for the position of Police Officer/Firefighter with the XYZ Police/Fire Department. Under California law, Government Code 1031(d) my prospective employer is required to conduct an investigation into my personal, medical, and psychological fitness to serve in this capacity.

I hereby authorize any Police Officer, Firefighter or any other authorized representative of the XYZ Police/Fire Department bearing this release, or a copy of it, within one year of its date, to obtain any information in your files pertaining to my employment, credit or educational records, including, but not limited to, academic achievement, attendance, athletic, personal history, performance evaluations, background investigations, polygraph examination results, any and all internal affairs investigations and disciplinary records including any files which are deemed to be confidential, sealed and/or secret, and any credit reports.

I also hereby authorize any Police Officer, Firefighter or other authorized representative of the XYZ Department bearing this release, or copy of it, within one year of its date, to obtain any medical records or medical information in the files of my current or former employer(s) or any current physician(s), or both, which pertain to my employment.

I hereby direct you to release this information upon request of the bearer. This release is executed with full knowledge and understanding that the information is for the official use of the XYZ Department.

Consent is granted for the XYZ Department to furnish the information described above to third parties in the course of fulfilling its official responsibilities. I further understand that I waive any right or opportunity to read or review any background investigation report prepared by or for the XYZ Department.

I hereby release you, as the custodian of such records, and any school, college, university or other educational institution, hospital or other repository of medical records, credit bureau, lending institution, consumer reporting agency, or retail business establishment including its officers, employees, or related personnel both individually and collectively, from any and all liability for damage of any kind, which may affect me, my heirs, family or associates because of compliance with this authorization.

Signature _____

Mandatory Documents
There are numerous documents you will need as the process continues. It would be best to make copies and organize them immediately. Inability to obtain even one piece of paper can hold up your process and make you look foolish to your background investigator. Most governmental agencies take a significant amount of time to process requests, so don't take any chances! Getting the following documentation and having it on file can give the edge you are looking for to be selected over the other candidates.

- Birth Certificate
  If you do not have one you will need to apply for a certified copy immediately at the county in which you were born. In most cases you can have them expedite the process for a small fee.

  If you were not born in the United States, you must have a Resident Alien Card or a Naturalization Certificate.

- Driver's License of the state in which you apply for peace officer status

- Proof of vehicle insurance

- Printout from DMV (10-year record); look for mistakes as they are not uncommon.

- Social Security Card
  Immediately contact the Social Security Office for a replacement card if you cannot locate it.

- High School Diploma
  Contact the school district to obtain a duplicate or certified copy. (Request copies of official and sealed transcripts)

- College Degree
  (Request copies of official and sealed transcripts)

- Personal Credit Report (including your credit score)
  (www.experian.com, 888-397-3742)

- Bankruptcy Filings
  (County Courthouse)

- Marriage License and/or Divorce Decree
  (County Courthouse)

- Military Discharge Papers (DD Form 214)
  www.archives.gov/veterans/evetrecs/

- Selective Service Registration (if male and born after 1/1/60)
  www.sss.gov, 847-688-6888

- Copies of work evaluations from past and present employers

- Police Academy Graduation Certificate (if applicable)

- P.O.S.T. and any other job-related certificates

Obtaining all of this documentation may seem overwhelming, but in reality with a few phone calls all of this can all be completed in a couple of days.

Polygraph Literally Means "Many Writings"

Almost all police departments and an ever increasing number of fire departments require a polygraph examination as part of the hiring process. This examination is conducted by an expert in the field of polygraph examinations who has completed hundreds of hours of on-site training. The examiner also attended mandatory seminars, required to maintain certification.

The American Polygraph Association, established in 1966, requires that current certified polygraph examiners read and evaluate the new examiner's charts to assist them in developing and maintaining proficiency. Polygraph examiners are meticulous, scrupulous and professional by nature. For more information, go to www.polygraph.org. This website includes a wide array of legal theories and frequently asked questions (FAQ).

Sergeant Matt Craig, an official polygraph examiner for police and fire applicants as well as criminal investigations, says the polygraph is an excellent tool for background investigators to discover the truth and resolve issues. Some departments conduct a polygraph examination first to determine if you are a suitable candidate for their agency; however, most departments complete a background investigation first, which will lead the polygraph examiner to explore specific areas of concern.

Sergeant Craig explains that the polygraph measures three different physiological changes: Cardiovascular response, respiratory response, and galvanic skin response. Your autonomic nervous system reacts to stress and displays your physiological response on a graph. The polygraph examiner will determine if you are being evasive or telling the truth.

The examiner will give you an opportunity to clear things up prior to the actual polygraph. He will have a conversation with you to ensure that you are comfortable, or as comfortable as you possibly can be in this situation. Once the electrodes are connected to your fingers and the blood pressure cuff is put on your arm, it's too late to clear anything up. I suggest you discuss any situations that you believe may affect you as soon as possible. It is easy to tell you not to be nervous, but if you have nothing to hide you truly shouldn't be nervous. The polygraph is very accurate and will assist the examiner in determining if you are truthful or not.

Your best option is to tell the entire truth about an incident you think will be a disqualifier, because in many cases it would not disqualify you. It never ceases to amaze us how applicants lie about what was just a slight bump in the road from a background investigator's point of view. An inconsequential incident that would not have disqualified the applicant now becomes an integrity issue, which will disqualify them forever!

At the conclusion of the examination, the examiner will key on specific areas where you showed a reaction. He will get more specific in the areas of concern and at this point knows you may not have been totally honest during the examination and/or background investigation. If there are logical reasons for a reaction and your explanations are believable, the examiner will test you again. Assuming you tell the truth, your polygraph will be concluded and your background investigator will be notified of the results. Depending on these results, you will continue in the process, be disqualified, or asked to withdraw.

166

Note: If it is suggested you withdraw at any time during the process, we strongly urge you do so. Background investigators rarely ask for this unless they already have enough negative information to disqualify you. If you argue or refuse to comply, you will receive a scathing letter of disqualification with the reasons you were disqualified placed in your file. Every agency that comes to review your file will see the letter and the chances of being hired decrease significantly. A withdrawal can help you eventually get hired in public safety if handled with dignity and a positive attitude.

Sergeant Craig offers several tips for a positive polygraph examination experience:

- Relax - don't be anxious - it really isn't that difficult
- Don't listen to others who have taken a polygraph; they give bad advice.
- Bc honest, disclose your indiscretions and free yourself from worry.
- Don't research "How to beat a poly;" it will just confuse you.

There are many websites that offer information on beating the test, for a small fee of course, or information that describes how to legally challenge the polygraph due to its history of inaccuracy and unprofessional operators.

My experience with polygraph examiners has been that they are meticulous, intelligent and extremely conscientious. I have seen enough tests, reactions and subsequent confessions to make me a believer. I will let you in on a little secret on how to beat the test…TELL THE TRUTH!

Psychological Testing

The Psychological Examination

The psychological evaluation for a law enforcement or fire position is not an evaluation of your mental health. Some personality types function better in public safety than others. The evaluation is designed to determine if your personality fits the public safety agency with which you are applying.

Just like the polygraph, the psychological examination is usually given near the end of the process and one of the last hurdles before you are hired.

Testing can vary depending on the department's philosophy of the test results as a tool. There are many different methods a psychologist uses to assess your suitability for public safety. Some psychologists will have the department proxy the written examination and then arrange for an appointment a few days later. It takes a couple of days to analyze the data and develop a specific plan for the applicant. The appointment usually takes about an hour in which you will be speaking with the doctor. Or, you may spend an entire day in their office completing tests while being observed by the doctor. He is watching how you deal with stress and how you cognitively process while under pressure.

Some psychologists put more of an emphasis on intelligence, while others place the emphasis on affability, common sense, and determination. Remember when dealing with the psychologist, wear a business suit unless specifically told not to and maintain a friendly but formal demeanor. Do not say anything off color just to be funny to lighten the mood during an awkward silence.

A psychological evaluation is designed for each specific position and is not an evaluation of your mental health. Below are some of the most common tests.

• **MMPI** - The MMPI also known as the Minnesota Multiphase Personality Inventory is the most often utilized test in public safety pre-employment testing. The test was developed by a psychologist and psychiatrist in the late 1930s. The MMPI was revised in 1989 and is a powerful instrument in the hands of an experienced psychologist. There arc almost 600 true and false questions in the test and it evaluates 10 clinical scales in distinguishing psychiatric disorders. This test has been used to determine suitability of parents during custody disputes and domestic violence situations. It has withstood numerous lawsuits and has always been victorious in the court room.

• **Sentence Completion** - There are no right or wrong answers as it is an evaluation of your common sense and deep rooted sentiments about a variety of subjects. A common question would be, "White people... you complete the sentence here."

• **16 Personality Factors** – 187 multiple choice questions in which you must utilize logic and intelligence. You will be mentally and physically exhausted after this test and the MMPI.

- **Autobiography** – This covers two areas - first to see if any deep rooted issues arise from your writing, and second to observe your writing ability and to view how your mind processes. Is it logical or does your written work ramble?

- **Wonderlic Test** – Used by the NFL to determine intelligence under pressure and has a major effect on whether players are drafted or not. Used by police and fire departments in the same manner. The test consists of 50 multiple choice questions with a 12-minute time limit. You receive a point for each correct answer. A score of 14 is equivalent to the mentality of an unskilled worker; a score of 21 indicates average intelligence. You will need a minimum score of 21 to qualify for police or fire, but successful public safety applicants are usually in the 25 range. Research the "Wonderlic Test" on the internet so you are not surprised by the formatting of the test.

- **California Psychological Inventory (CPI)** – A personality questionnaire used to assess temperament, managerial potential and tough mindedness, which ultimately helps reduce selection errors.

- **Wecshler Adult Intelligence Scale (WAIS)** – Developed in 1955 and consists of 11 separate tests, including general knowledge, math and vocabulary.

- **California Personality Inventory (CPI)** – Originally developed for business executives in 1953, it is a commonly used test for public safety applicants.

- **Psychologist's Interview** – The psychologist analyzes data and conducts a one on one interview to determine suitability for public safety.

Psychologist Interview

I interviewed Dr. Eric Gruver who has been the screening Psychologist for numerous law enforcement and fire agencies since 1977 for both entry level and lateral officers. Dr. Gruver is also a leading expert in the field of critical incident debriefings. He has interviewed thousands of prospective applicants and makes recommendations based on his findings.

Dr. Gruver, like other psychologists, assesses the probability of success for a candidate for each particular police or fire agency. They are screening for mental stability, personal disorders and psychosis issues. Each agency and psychologist develops a critical criteria checklist and determines what the essential attributes are for their particular department. It is possible to be qualified for one agency and not qualified for another. A passive, more cerebral, applicant may work well in one city and be in danger in a city with a higher level of activity.

Some factors are: Does the candidate fit the personality of the department? Are they aggressive enough for a respective department or too aggressive for another?

Dr. Gruver stated it is normal to be nervous and anxious during the exam, but realize if you have gotten this far in the process, you are a good candidate. He said the candidate should not try to "fake it," by trying to manipulate the tests. Answer truthfully and understand that he has access to and will be familiar with your background investigation. If you insist on being somebody you're not, results will be skewed as the tests have built-in validity scales. Your test will be ruled invalid and will not bode well for your examination results.

Dr. Gruver's advice is to understand that the psychological examination may be an extensive and grueling process and for the candidates to "be themselves." Be direct and honest about your mistakes and do not display evasiveness. You may or may not be disqualified by something you did in your past, but psychologists do understand that people change and if you have shown maturity and now handle yourself with integrity, common sense and good judgment, you will have a good chance of getting hired for a position in public safety.

Dr. Gruver is available for consultation at (714) 544-4434. He may also be reached via mail at, 17772, 17th Street, Suite 106, Tustin, CA 92680.

Notification of Whether You Passed or Failed

It is highly unlikely that a psychologist will advise an applicant of the status of the evaluation. The doctor most likely already knows whether or not you passed when you leave his office, but still needs time to analyze and review the data. On occasion, a discussion with the background investigator after the psychological interview is necessary, which may point out discrepancies or new information that needs to be re-evaluated. Most applicants have a sense if the evaluation went well. If the doctor spends an inordinate amount of time on your drinking pattern or unstable job history, that is not a good sign.

Medical Examination
Americans with Disabilities Act

There are several aspects to a medical examination for public safety. The department is permitted by law to have the applicant perform tests relating to "essential job functions." When testing and determining if the candidate is suitable for employment, the department must stay within the guidelines of the "Americans with Disabilities Act." This states that as long as the applicant's limitation does not create an undue hardship on the employer, they would be required to make "reasonable accommodations."

An example of a reasonable accommodation would be: If diabetes were discovered, would the applicant still be able to do the job? If the employer allowed a few minutes every six hours for the applicant to administer himself an injection of insulin, does it create undue hardship to the employer? Now let's say the diabetic employee will need five minutes every hour to rest, and in addition needs a 30-minute nap every four hours. Would this situation create an undue hardship for the employer? What "reasonable" means is very subjective and is occasionally interpreted by the court.

Medical Examination

This is usually the last component of the process. Doctors will measure cardiovascular health by conducting a stress test on the treadmill. There will be electrodes hooked up to your chest and your blood pressure will be monitored. If you haven't started running yet, you should start immediately. If you are a smoker you should have stopped yesterday. Smoking affects your overall cardiovascular health that could eliminate you from the hiring process. As stated earlier many agencies require you to sign a non-tobacco use agreement prior to obtaining a position with their agency. Violation of this contract could result in termination; the logic being non-smokers have less medical problems than smokers, consequently requiring less money to insure them, both on and off the job.

You will be given chest X-rays and receive comprehensive blood results from a phlebotomy laboratory. You will be asked questions about night blindness and being color blind. If anything medical keeps you from attaining your objective don't give up! Find a way! Seek a second and third opinion if necessary. There may be a method to correct your specific deficiency.

Color blindness disqualifies numerous applicants and shouldn't as there are now correctable lenses available. Lasik surgery is an option for certain types of visual acuity problems.

Specialists

It is not uncommon to have a hold-up in your medical examination. Don't panic! It's probably not as big a deal as you think. Due to our litigious society, no one is willing to make a decision and doctors are no exception. They believe, and rightfully so, that the parasitic lawyers will sue a well-intentioned doctor at the slightest opportunity of a pay day. Consequently, an insignificant medical condition such as a heart murmur, irregular heart beat, and other common ailments will be referred to a specialist for a final decision. It is in your best interest to find a specialist you have a rapport with. The specialist will communicate with the City doctor and will complete the paperwork allowing you to continue in the process.

You will be given a drug test. If you have to worry about this, we made a mistake letting you get this far!

How to Act When Awaiting an Academy Appointment

A friend of mine was at a party and everyone in attendance was aware he would be starting the police academy within a couple of weeks. A supposed friend of his looked directly at him and lit up a joint. First of all, that person is not a friend as he displayed blatant disrespect. The smoker was probably testing my friend to see how he would react when being placed in a bad position. Do not let yourself be placed in that type of predicament; people will be jealous and test you in many ways. It is just a sign of many things to come that will test your character as you progress through your career. Be careful who you associate with, as one transgression may alter your future.

Stress or Non-Stress Academy

There are generally two types of academies: Stress and non-stress. A stress academy involves, time constraints, yelling, screaming and demeaning the recruits while applying continuous pressure and stress the duration of the academy. They will give you homework that will keep you up most of the night, causing sleep deprivation. The academy is a test to determine if you can handle the streets.

The positive aspect of a stress academy is that nothing will bother you after you have completed the program and you will have better control of your emotions. If I had a choice to select one of two applicants to be my partner - one graduating from a stress academy and the other from a non-stress academy - - assuming I had no other knowledge to go on, I would choose the stress academy graduate. The stress academy does have a weakness however. It does not create an environment conducive to academics. The basics will be memorized, but learning shuts down during periods of high stress levels. While in class the recruit goes into survival mode, mentally rejecting the information being taught. In our opinion, the non-stress academies graduate a higher level of academic recruit.

It is very difficult to graduate from a stress academy unless you are sponsored by a department. Stress academies are intense and require a built-in support group that a non-sponsored recruit would not have. If you are a non-sponsored recruit and decide to attend a stress academy and graduate near or at the top of your class, you can write your own ticket. Historically though, a non-sponsored recruit graduating is rare and we don't recommend it.

A non-stress academy is similar to a college course, only all day and all week. There is limited stress involved, which enhances learning and retention. The argument for attending a non-stress academy is that recruits are sent to their departments with a higher base knowledge. The argument against a department sending their recruit to a non-stress academy is that the recruit has not been tested.

After conducting an informal survey of veteran officers, it appears there is no significant difference in the performance of an officer, whether or not he or she attended a stress or non stress academy.

Extended Academy

For you working men and women, another option is attending what is called an "extended academy." These are becoming more popular as recruits can still support themselves and their family while attending in the evening and on the weekends.

If you are worried that this type of academy doesn't hold as much clout as the others, put that out of your mind right now. You will fulfill the same requirements and be trained by professionals just as you would the full-time academy and be just as marketable. In fact one of my best officers graduated from an extended academy.

Academy

There are two ways to go through an academy, sponsored and un-sponsored. Sponsored is when a department is paying you while you are attending the academy. This is the way most police departments operate; it will be difficult for you to attend unsponsored as it may appear to departments you have had trouble getting hired. Fire has only 25% of the job openings law enforcement has and enjoys a type of supply and demand. It is rare to be sponsored through a fire academy. You must pay your own way while you somehow make a living, and in most cases you are not allowed to have full-time employment while attending the academy.

Each individual police or fire agency has their own outlook regarding sponsoring recruits and it is determined by many variables; mainly budget and the availability of qualified applicants. The higher the number of qualified applicants in the applicant pool reduces their need to pay you while you train. You will also notice a trend by geographic region.

There are both pros and cons to being sponsored and unsponsored. In being sponsored you are an at-will employee. If you have a bad week or even a bad day, there is a very real risk of being fired. Going unsponsored has the benefit of having your overall performance judged rather than a few independent situations. You may also be taken under the wing of an academy staff that sees potential in you. The academy staff is comprised of professionals who are currently working in police or fire or have been sworn in the past. Their opinion is respected and we often rely on feedback from them as they know who belongs in public safety and who does not.

One particular recruit was not hired because we believed he wasn't tough enough to handle the streets. He decided to go to the academy un-sponsored and was in the top five by the second half of the program. A tactical officer from the academy was a strong proponent of this recruit and based on our respect for him we hired this recruit. We have never regretted the decision as he is doing well and is a very good officer. What I am saying is to go all out and impress your instructors every second of the day. They do have input into your future, and if you prove to them you belong, you stand a good chance of being hired!

Never Quit

If you are destined for public safety, nothing should deter you. If you make a tenacious effort, the percentages of landing your dream job are in your favor. When I hear of someone giving up, I know the job wasn't in their heart in the first place and probably indirectly contributed to their inability to obtain a position. Never, ever give up!

Conclusion

We have tried to educate you on the various law enforcement and fire agencies, as well as the many types of hiring processes. Hopefully you have a better understanding of the basics of applications, oral interviews and the background investigation. If you give your best effort and pay attention to the advice in this book you should do well. Good luck and welcome to the family!

It is time to get to it! Began researching any law enforcement, criminal justice or fire technology program, and decide which one is best for you and follow your dream!

Both of the authors are available for public speaking engagements throughout the country. For further information contact either one directly at their Email address.

We love to hear from you. Send your questions and success stories/testimonials to us:
Email Steve Winston at ihelppolice@yahoo.com.
Email Pete Bollinger at bolly7@aol.com.

If you are still uncertain of the specific area of law enforcement you should enter, speak with your academic counselor and/or instructors. They want what is best for you and your success will be their success.

You may also visit our secured website (www.hiredbypolice.com or www.policeandfirepublishing.com) to add your name to our mailing list. As police officers and firefighters, we value privacy above all else. Your personal information will be confidential and will not be released, sold or loaned to any other entity.

Never stop learning! Read and study every book in this series and you will be exponentially more prepared than you would be with any other method.

Advanced Report Writing Workbook
Background Investigation Workbook
CSI: Beyond the Yellow Tape Textbook W/instructional DVD
Family Violence: Beyond the Bruises
Gang Life
How to Get Hired in Law Enforcement
How to Get Hired in Probation and Parole
How to Pass the Oral Interview W/instructional DVD
Law Enforcement Written Test Workbook

To order, contact hiredbypolice.com or policeandfirepublishing.com. To order in volume, contact Pete Bollinger at 951-237-3351 or bolly7@aol.com.